PROJECT
VANGUARD

THE NASA HISTORY

Constance McLaughlin Green • Milton Lomask

Introduction to the Dover edition by Paul Dickson
Foreword by Charles A. Lindbergh

Dover Publications, Inc.
Mineola, New York

Bibliographical Note

This Dover edition, first published in 2009, is an unabridged and slightly corrected republication of the 1971 Smithsonian Institution Press edition of the work first published in 1970 in the NASA Historical Series as NASA SP-4202 under the title *Vanguard: A History.* A new introduction by Paul Dickson has been added to this edition.

Library of Congress Cataloging-in-Publication Data

Green, Constance McLaughlin, 1897–1975.
    [Vanguard]
    Project Vanguard : the NASA history / Constance McLaughlin Green, Milton Lomask ; foreword by Charles A. Lindbergh. — Dover ed. / introduction to the Dover edition by Paul Dickson.
        p. cm.
    Originally published: Vanguard. Washington, D.C. : Smithsonian Institution Press, 1971.
    ISBN-13: 978-0-486-46755-9
    ISBN-10: 0-486-46755-4
    1. Project Vanguard. 2. United States. National Aeronautics and Space Administration. I. Lomask, Milton. II. Title.

TL796.5.U6V35 2009
629.430973—dc22

                                     2008038043

Manufactured in the United States of America
Dover Publications, Inc., 31 East 2nd Street, Mineola, N.Y. 11501

INTRODUCTION TO THE DOVER EDITION

ON MARCH 17, 2008 a barely-noticed anniversary took place with nary a mention in the mainstream news media. The softball-sized Vanguard 1 satellite marked its 50th year in orbit above Earth making it, among other things, the oldest manmade object in space. Vanguard I had circled the planet almost 200,000 times—about six billion miles—on an orbit traveling from 400 to 2,400 miles above the Earth's surface. Estimates of its orbital lifetime range from 240 to 2000 years.

Launched into orbit at 7:15 a.m. on a clear St. Patrick's Day morning five months and thirteen days after the launch of Sputnik 1 by the Soviet Union, and some six weeks after America's first successful orbital shot, Explorer 1, Vanguard 1 was followed in 1959 by Vanguard 2 and 3. Scientific results from this series of Earth-circling satellites included the first geodetic studies indicating the Earth's slightly "pear" shape, a survey of the Earth's magnetic field, the location of the lower edge of the Earth's radiation belts, and a count of micrometeorite impacts.

The Vanguard satellite pioneered the use of solid state devices, printed circuits and the general principle of miniaturization. The idea was to create as much scientific power as possible within the smallest space possible, but this was not generally understood in 1958 when a computer was still an object which needed to be housed in a large air-conditioned room. It was the first satellite to use solar cells to create power. There were two transmitters aboard the satellite: one powered by conventional batteries the other using solar cells. The solar-powered transmitter was simplicity itself—thin layers of silicon which when exposed to sunlight absorbed the sun's energy and transmitted a small electric current, a mere 0.06 watts, but sufficient to activate a radio beam. The satellite was also placed into almost perfect orbit far above the denser part of the atmosphere where there is little atmospheric drag– the reason why it will remain in orbit for centuries.

Although Vanguard 1's radio went dead in 1964 and Vanguard no longer transmits information, it is a reminder of the fact that the Vanguard program was really a major success despite a spectacular and widely televised failure on the launch pad at the very beginning of the space race when the United States, rushing to respond to the launch of Sputnik on October 4, 1957, attempted to launch Vanguard later in the same year.

At precisely 11:44.55 a.m. on Friday, December 6, 1957, with the whole world watching, the slender vehicle rose a few feet off the launch platform, shuddered slightly, buckled under its own weight, burst into flames, and collapsed. "Lookout! Oh God, no!" were the first words heard in the blockhouse. The Cape Canaveral "launch" had lasted a mere two seconds before turning into a spectacular fireball. Its tiny 3.2 pound payload was thrown free of the fire, rolled into hiding in the scrub brush and then started beeping. A festive group of reporters watching through binoculars from the balcony of the new Vanguard Motel nearby were horrified. The late Ross Mark of London's *Daily Express* watched as white smoke turned black and then "everything burst into flames." Columnist Dorothy Kilgallen asked, "Why doesn't somebody go out there, find it and kill it?"

But this was just the beginning. On the morning after the failure, at United Nations headquarters in New York City, the Soviet delegation formally offered financial aid to the United States as part of a program of technical assistance to backward nations. The humiliation was most complete at this point. The Russians played this for all it is worth—every chance they got for weeks they used the Sputniks to tell the world how inadequate America was.

Now, it seemed, America had proven the point.

Unfortunately, what the press, politicians and public had failed to realize was that the Vanguard team knew they were dealing with a new and untested rocket and failure at this stage was not unexpected. Largely forgotten was the fact that three-stage rocket was designated TV-3, for Test Vehicle 3, and it had been originally scheduled to be just that, a test, but that "test" was moved up several months, and made into a full-fledged attempt at a satellite launch as a reply to Sputnik.

Yet, the highly publicized and telegenic Vanguard failure on December 6, 1957, forever created a lasting negative impression of a spectacular and public propaganda catastrophe. The film of that first sensational explosion has become part of virtually every television documentary on the Space Race or the Cold War. When John Hagen passed away in 1990, the obituary writers remembered him more for his failure than his success. *The New York Times* called him "...a leading figure in the dramatic and disappointing early days of the American space program." After recounting his failures we hear of the 3.2 pound satellite which is the oldest manmade object in space, but there is no mention of its many successes. The errors are not of fact but of omission. For years, critics dwelt on its price tag. "An expensive fiasco for the most part" is how Isaac Asimov later dismissed the Vanguard program.

Historic inaccuracy often prevails over the subtleties of what really happened and the blurred conventional wisdom today sees Vanguard as metaphoric of failure.

Constance McLaughlin Green and Milton Lomask were able to tell the real story of Vanguard—the one that is so often obscured and misrepresented—in their 1969 narrative which is now republished as a Dover edition complete with the original foreword by Charles A. Lindbergh. *Vanguard, A History* was sponsored by NASA which commissioned skilled writers to tell this story which is both accurate, academically correct and accessible to the general reader.

This is also the first work in a series of space histories to be published by Dover in the coming months and years in cooperation with the NASA History Office.

—Paul Dickson

FOREWORD

As a youth, near the beginning of this 20th century, Robert Hutchins Goddard began to dream of spatial voyages, thereafter becoming the great pioneer of astronautics. The Vanguard project, which this book records, was a major step in transforming his dreams to reality. It may be said to have rooted into Goddard's New Mexico rocket launchings, and to have stretched far upward toward man's landing on the moon.

This is a record of amazing human accomplishment. It is also a record of conflicting values, policies, and ideas, from which we have much to learn. It shows how easily an admirable framework of scientific success can be screened from public view by a fictitious coating of failure; yet how important that coating is in its effect on world psychology and national prestige.

Herein is portrayed both the genius and the ineptness of our American way of life. On one hand, Vanguard history rests proudly with outstanding accomplishments; on the other, it emphasizes clearly how much more could have been accomplished through inter-service cooperation and support that was withheld. One is made aware of the penalties brought by such diverse elements as extended wartime hatred and a sensation-seeking press. Even in retrospect, we view Project Vanguard through the haze of an environment that was out of its control—an environment including atomic weapons, *Sputnik,* and cold war with the Soviet Union.

These chapters bring out again the age-old conflict that continues between security and progress, in spite of their relationship. My own first contact with the Vanguard program was a part of this conflict. It exemplifies the obstacles, sometimes unavoidable, that delayed the launching of America's Number I satellite.

In 1955, I was sitting as a member of one of the ballistic-missile scientific committees when the subject of orbiting a satellite was broached. Everyone present agreed that a satellite-launching project was practical, important, and deserving of support. But what priority should be assigned to it? We were working under staggering concepts of nuclear warfare, and the imperative need to prevent a strike against the United States by the Soviet Union. Our mission was to speed the development of intercontinental ballistic missiles with sufficient retaliatory power to discourage attack. Members of the committee believed that the security of our nation and civilization would depend on our ability to shoot across oceans quickly and accurately by the time Russia had missiles available for shooting at us.

A shortage of scientists, engineers, and facilities existed in fields of missiles and space. Military projects needed every man available. The consensus of committee opinion was that we should concentrate on security requirements and assign to the Vanguard program a secondary place. There would be time to orbit satellites after our nuclear-warhead missiles were perfected and adequate marksmanship achieved.

My second contact with American satellite development relates largely to the realm of "might have been." It impinged only indirectly on Project Vanguard; but it throws light on our apparent early space-backwardness in relation to the Soviet Union. During my visit to the Army's Redstone Arsenal at Huntsville on a military mission, Wernher von Braun showed me one of his mock-ups and told me of his plans for orbiting a satellite. He had been unable to obtain the authorizations and funds required to put an actual satellite in orbit; but according to his estimates, he could have done so many months before Russia's *Sputnik I* was launched—at a minute cost as measured by present-day space expenditures.

Why was the Redstone-von Braun satellite project not supported? Answers vary with the person talked to: The Navy's brilliant developments in satellite instrumentation had tipped the choice to Vanguard, and budgetary restrictions had prevented a paralleling project. The name Redstone was too closely associated with military missiles. Vanguard offered lower costs, more growth potential, longer duration of orbiting. We would eventually gain more scientific information through Vanguard than through Redstone. To these observations, I can add from my own experience that inter-service rivalry exerted strong influence; also, that any conclusion drawn would be incomplete without taking into account the antagonism still existing toward von Braun and his co-workers because of their service on the German side of World War II.

My third contact with the satellite program was one nearly every American experienced: the effect of a superficial and sensational press. This has retarded the development of many important projects, Vanguard among them. It places such an unrealistic penalty on developmental errors that increased costs and delays result through excessive attempts to avoid them. Generally speaking, success comes faster when a percentage of error is not only accepted but included in planning schedules.

In addition to the over-caution caused by press and public monitoring, the emphasis of scientific perfection had its retarding effect on our launching of a satellite. How often I heard the statement that only the quality of data obtained mattered, and that the United States should not take part in a race with the Soviet Union to orbit a few pounds of matter around the earth. In spite of frequent warnings both from within and without scientific circles, an ivory tower of intellect was built—and suddenly shattered by *Sputnik*.

The amazing effectiveness of *Sputnik* was due even more to the American reaction than to the Soviet satellite. That the Russians are a great people has been proven beyond doubt. Over and over again through generations, they have made outstanding contributions to fields of scientific progress. One of these fields is astronautics. From Tsiolkovskiy through Gagarin to scientists, engineers, and astronauts of current days, Russians have been acclaimed throughout the world. The launching of *Sputnik I* was another great Russian accomplishment, and it was proper that we regarded it as such. One might have expected Washington to cable congratulations to Moscow, possibly to add that we would soon complement with our own satellites this Russian contribution to the International Geophysical Year, and then to proceed with the carefully considered plans we had adopted.

Instead, an atmosphere bordering on hysteria pervaded the United States. After having decided that we were not primarily in a race to launch the world's first satellite, we seemed to start running, laps behind, in a race already won, and to convince ourselves that we had been in it right along. In convincing ourselves, we went far toward convincing the rest of the world along with us—with a resulting lowering of American prestige.

Why did America, starting with the lead that Robert Goddard gave us, permit another country to be first in orbiting a satellite? I think we would gain more basic enlightenment from asking why we were so disturbed psychologically by the orbiting of *Sputnik*. We led the world in science, technology, and military power, even in the development of astronautics scientifically. Our leading minds had suppressed the spectacular in space in order to emphasize the scientific. Yet when the spectacular appeared as a pinpoint of light above our heads, it seemed to obliterate our great accomplishments of science. From such an irrational reaction, what rational conclusions can we draw?

That the spectacular and psychological exert tremendous influence on life is obvious. That we underestimated this influence in planning our early satellite program is just as clear. That, in hindsight, we should have changed the relative priority between missiles and satellites can still be argued. That we missed an opportunity to have held the established priority and still orbited the world's first satellite seems highly probable.

In rational analysis, two facts emerge to silhouette against a turbulent background. The Project Vanguard objective was accomplished during the International Geophysical Year, as planned, and our military forces achieved a retaliatory power before the Soviet Union could have destroyed our civilization by a ballistic-missile attack. Both programs were successful despite the many obstacles encountered.

This carefully researched history of Project Vanguard, resulting from years of study by Constance McLaughlin Green and Milton Lomask, escorts the reader step by step through planning, setbacks, successes, and final launchings, including effects of individual and mass psychology. Since contributions by the armed services form the web on which the following chapters pattern, it seems appropriate to apply to the authors, as well as to Vanguard, those high compliments of military terminology: *Mission accomplished* and *well done.*

To this decade-later evaluation, I believe it is pertinent to add that Project Vanguard contributed in major ways to the manned lunar orbitings and landings in which principles of scientific perfection were maintained and America was first.

Charles A. Lindbergh

11 August 1969

CONTENTS

LIST OF ILLUSTRATIONS

PREFACE

THIS BOOK deals with the origin, course of development, and results of the first American earth satellite project, one of several programs planned for the International Geophysical Year. Primarily an analysis of the scientific and technical problems in this pioneering venture in the exploration of outer space, the text also examines the organization of an undertaking bound by an inexorably fixed time limit, discusses briefly the climate of American opinion both before and after the launchings of the first Russian Sputniks, and concludes with a somewhat cursory evaluation of what the satellite program contributed to human knowledge.

Written in lay language insofar as the authors could translate scientific and technical terms into everyday English, the book nevertheless is not one for casual reading. The very multiplicity of Federal agencies, quasi-governmental bodies, and private organizations that shared in the project complicates the story. Indeed in some degree the interrelationships of these groups and key individuals within them constitute a central theme of this study. Even so, by no means all the several hundred people whose dedicated work made satellite flights possible are mentioned by name in the text. To have identified each person and explained his role would have turned this book into a large tome.

As authors we divided our responsibilities by topic: Milton Lomask wrote the chapters on the field tests and satellite launchings at Cape Canaveral, Constance McLaughlin Green the rest. Jointly and separately we are deeply indebted to the late Alan T. Waterman, the late Lloyd V. Berkner, Hugh Odishaw, Homer E. Newell, John P. Hagen, Milton Rosen, and a score of other men for their help in sorting out the pieces of a complex tale and for drawing upon their memories or private papers for information nowhere officially recorded. We are equally grateful to Charles A. Lindbergh for writing an introduction that puts the material into the perspective of the mid-1950s. NASA historians Eugene M. Emme and Frank W. Anderson assisted us at many points, and Nancy L. Ebert of the NASA Scientific and Technical Information Division devoted many hours of overtime to typing the text. We want to thank also the men who subjected the manuscript in whole or in part to a critical reading.

Time has modified many judgments about Project Vanguard, but a good deal of the story is still controversial. While we have endeavored to present it dispassionately, doubtless few of the participants will subscribe to our every interpretation. Scientists, engineers, technicians, industrial managers, the President and the White House staff, the Bureau of the Budget, and members of Congress tended to view the program as it progressed from markedly different standpoints. In the final analysis the verdict of our readers may determine the place of Vanguard in history.

Constance McLaughlin Green
Milton Lomask

Washington, D.C.
September 1969

1
BACKGROUND OF SPACE EXPLORATION

PEOPLE the world over speak of the "Space Age" as beginning with the launching of the Russian Sputnik on 4 October 1957. Yet Americans might well set the date back at least to July 1955 when the White House, through President Eisenhower's press secretary, announced that the United States planned to launch a man-made earth satellite as an American contribution to the International Geophysical Year. If the undertaking seemed bizarre to much of the American public at that time, to astrophysicists and some of the military the government's decision was a source of elation: after years of waiting they had won official support for a project that promised to provide an invaluable tool for basic research in the regions beyond the upper atmosphere. Six weeks later, after a statement came from the Pentagon that the Navy was to take charge of the launching program, most Americans apparently forgot about it. It would not again assume great importance until October 1957.

Every major scientific advance has depended upon two basic elements, first, imaginative perception and, second, continually refined tools to observe, measure, and record phenomena that support, alter, or demolish a tentative hypothesis. This process of basic research often seems to have no immediate utility, but, as one scientist pointed out in 1957, it took Samuel Langley's and the Wright brothers' experiments in aerodynamics to make human flight possible, and Hans Bethe's abstruse calculations on the nature of the sun's energy led to the birth of the hydrogen bomb, just as Isaac Newton's laws of gravity, motion, and thermodynamics furnished the principles upon the application of which the exploration of outer space began and is proceeding. In space exploration the data fed back to scientists from instrumented satellites have been of utmost importance. The continuing improvement of such research tools opens up the prospect of greatly enlarging knowledge of the world we live in and making new applications of that knowledge.

In the decade before Sputnik, however, laymen tended to ridicule the idea of putting a man-made object into orbit about the earth. Even if the feat were possible, what purpose would it serve except to show that it could be done? As early as 1903, to be sure, Konstantin Tsiolkovskiy, a Russian scientist, had proved mathematically the feasibility of using the reactive force that lifts a rocket to eject a vehicle into space above the pull of the earth's gravity. Twenty years later Romanian-born Hermann Oberth had independently worked out similar formulas, but before the 1950s, outside a very small circle of rocket buffs, the studies of both men remained virtually unknown in the English-speaking world. Neither had built a usable rocket to demonstrate the validity of his theories, and, preoccupied as each was with plans for human journeys to the moon and planets, neither had so much as mentioned an unmanned artificial satellite.[1] Indeed until communication by means of radio waves had developed far beyond the techniques of the 1930s and early 1940s, the launching of an inanimate body into the heavens could have little appeal for either the scientist or the romantic dreamer. And in mid-century only a handful of men were fully aware of the potentialities of telemetry.[2]

Of greater importance to the future of space exploration than the theoretical studies of the two European mathematicians was the work of the American physicist, Robert Goddard. While engaged in post-graduate work at Princeton University before World War I, Goddard had demonstrated in the laboratory that rocket propulsion would function in a vacuum, and in 1917 he received a grant of $5,000 from the Smithsonian Institution to continue his experiments. Under this grant the Smithsonian published his report of his theory and early experiments, *Method of Reaching Extreme Altitudes.* In 1918 he had successfully developed a solid-fuel ballistic rocket in which, however, even the United States Army lost interest after the Armistice. Convinced that rockets would eventually permit travel into outer space, Goddard after the war had continued his research at Clark University, seeking to develop vehicles that could penetrate into the ionosphere. In contrast to Tsiolkovskiy and Oberth, he set himself to devising practical means of attaining the goal they all three aspired to. In 1926 he successfully launched a rocket propelled by gasoline and liquid oxygen, a "first" that ranks in fame with the Wright brothers' Kitty Hawk flights of 1903. With the help of Charles Lindbergh after his dramatic solo transatlantic flight, Goddard obtained a grant of $50,000 from Daniel Guggenheim and equipped a small laboratory in New Mexico where he built several rockets. In 1937, assisted by grants from the Daniel and Florence Guggenheim Foundation, he launched a rocket that reached an altitude of

At Roswell, New Mexico, Robert H. Goddard and colleagues examine
rocket components after a successful flight on 19 May 1937.
(Photo courtesy of Mrs. Robert H. Goddard)

9,000 feet. Although not many people in the United States knew much about his work, a few had followed it as closely as his secretiveness allowed them to; among them were members of the American Interplanetary Space Society, organized in 1930 and later renamed the American Rocket Society. With the coming of World War II Goddard abandoned his field experiments, but the Navy employed him to help in developing liquid propellants for JATO, that is, jet-assisted takeoff for aircraft. When the Nazi "buzz" bombs of 1943 and the supersonic "Vengeance" missiles—the "V–2s" that rained on London during 1944 and early 1945—awakened the entire world to the potentialities of rockets as weapons, a good many physicists and military men studied his findings with attention. By a twist of fate, Goddard, who was even more interested in astronautics than in weaponry, died in 1945, fourteen years before most of his countrymen acknowledged manned space exploration as feasible and recognized his basic contribution to it by naming the government's new multi-million-dollar experimental station at Beltsville, Maryland, "The Goddard Space Flight Center." [3]

During 1943 and early 1944, Commander Harvey Hall, Lloyd Berkner, and several other scientists in Navy service examined the chances of the Nazis' making such advances in rocketry that they could put earth satellites into orbit either for reconnaissance or for relaying what scare pieces in the press called "death rays." While the investigators foresaw well before the first V–2 struck Britain that German experts could build rockets capable of reaching targets a few hundred miles distant, study showed that the state of the art was not yet at a stage to overcome the engineering difficulties of firing a rocket to a sufficient altitude to launch a body into the ionosphere, the region between 50 and 250 miles above the earth's surface. In the process of arriving at that conclusion members of the intelligence team, like Tsiolkovskiy and Oberth before them, worked out the mathematical formulas of the velocities needed. Once technology had progressed further, these men knew, an artificial earth-circling satellite would be entirely feasible. More important, if it were equipped with a transmitter and recording devices, it would provide an invaluable means of obtaining information about outer space.[4]

At the end of the war, when most Americans wanted to forget about rockets and everything military, these men were eager to pursue rocket development in order to further scientific research. In 1888 Simon Newcomb, the most eminent American astronomer of his day, had declared: "We are probably nearing the limit of all we can know about astronomy." In 1945, despite powerful new telescopes and notable advances in radio techniques, that pronouncement appeared still true unless observations made above the earth's atmosphere were to become possible. Only a mighty rocket could reach beyond the blanket of the earth's atmosphere; and in the United States only the armed services possessed the means of procuring rockets with sufficient thrust to attain the necessary altitude. At the same time a number of officers wanted to experiment with improving rockets as weapons. Each group followed a somewhat different course during the next few years, but each gave some thought to launching an "earth-circling spaceship," since, irrespective of ultimate purpose, the requirements for launching and flight control were similar. The character of those tentative early plans bears examination, if only because of the consequences of their rejection.

"Operation Paperclip," the first official Army project aimed at acquiring German know-how about rocketry and technology, grew out of the capture of a hundred of the notorious V–2s and out of interrogations of key scientists and engineers who had worked at the Nazi's rocket research and development base at Peenemuende. Hence the decision to bring to the United States about one hundred twenty of the German experts along with the captured missiles and spare parts. Before the arrival of the Germans, General Donald Putt of the Army Air Forces outlined to officers at Wright Field some of the Nazi schemes for putting space platforms into the ionosphere; when his listeners laughed at what appeared to be a tall tale, he assured them that these were far from silly vaporings and were likely to materialize before the end of the century. Still the haughtiness of the Germans who landed at Wright Field in the autumn of 1945 was not endearing to the Americans who had to work with them. The Navy wanted none of them, whatever their skills. During a searching interrogation before the group left Germany a former German general had remarked testily that had Hitler not been so pig-headed the Nazi team might now be giving orders to American engineers; to which the American scientist conducting the questioning growled in reply that Americans would never have permitted a Hitler to rise to power.[5]

At the Army Ordnance Proving Ground at White Sands in the desert country of southern New Mexico, German technicians, however, worked along with American officers and field crews in putting reassembled V–2s to use for research. As replacing the explosive in the warhead with scientific instruments and ballast would permit observing and recording data on the upper atmosphere, the Army invited other government agencies and universities to share in making high-altitude measurements by this means. Assisted by the German rocketeers headed by Wernher von Braun, the General Electric Company under a contract with the Army took charge of the launchings. Scientists from the five participating universities and from laboratories of the armed services designed and built the instruments placed in the rockets' noses. In the course of the next five years teams from each of the three military services and the universities assembled information from successful launchings of forty instrumented V–2s. In June 1946 a V–2, the first probe using instruments devised by members of the newly organized Rocket Sonde Research Section of the Naval Research Laboratory, carried to an altitude of sixty-seven miles a Geiger-counter telescope to detect cosmic rays, pressure and temperature gauges, a spectrograph, and radio transmitters. During January and February 1946 NRL scientists had investigated the possibility of launching an instrumented earth satellite in this fashion, only to conclude reluctantly that engineering techniques were still too unsophisticated to make it practical; for the time being, the Laboratory would gain more by perfecting instruments to be emplaced in and recovered from V–2s. As successive shots set higher altitude records, new spectroscopic equipment developed by the Micron Waves Branch of the Laboratory's Optics Division produced a number of excellent ultraviolet and x-ray spectra, measured night air glow, and determined ozone concentration.[6] In the interim the Army's "Bumper" project produced and successfully flew a two-stage rocket consisting of a "WAC Corporal" missile superimposed on a V–2.

After each launching, an unofficial volunteer panel of scientists and technicians, soon known as the Upper Atmosphere Rocket Research Panel, discussed the findings. Indeed the panel coordinated and guided the research that built up a considerable body of data on the nature of the upper atmosphere. Nevertheless, because the supply of V–2s would not last indefinitely, and because a rocket built expressly for research would have distinct advantages, the NRL staff early decided to draw up specifications for a new sounding rocket. Although the Applied Physics Laboratory of the Johns Hopkins University, under contract with the Navy's Bureau of Ordnance and the Office of Naval Research, was modifying the "WAC Corporal" to develop the fin-stabilized Aerobee research rocket, NRL wanted a model with a sensitive steering mechanism and gyroscopic controls. In August 1946 the Glenn L. Martin Company won the contract to design and construct a vehicle that would meet the NRL requirements.[7]

Four months before the Army Ordnance department started work on captured V–2s, the Navy Bureau of Aeronautics had initiated a more ambitious research scheme with the appointment of a Committee for Evaluating the Feasibility of Space Rocketry. Unmistakably inspired by the ideas of members of the Navy intelligence team which had investigated Nazi capabilities in rocketry during the war, and, like that earlier group, directed by the brilliant Harvey Hall, the committee embarked upon an intensive study of the physical requirements and the technical resources available for launching a vessel into orbit about the earth. By 22 October 1945, the committee had drafted recommendations urging the Bureau of Aeronautics to sponsor an experimental program to devise an earth-orbiting "space ship"

6

launched by a single-stage rocket, propelled by liquid hydrogen and liquid oxygen, and carrying electronic equipment that could collect and transmit back to earth scientific information about the upper atmosphere. Here was a revolutionary proposal. If based on the speculative thinking of Navy scientists in 1944, it was now fortified by careful computations. Designed solely for research, the unmanned instrumented satellite weighing about two thousand pounds and put into orbit by a rocket motor burning a new type of fuel should be able to stay aloft for days instead of the seconds possible with vertical probing rockets. Nazi experts at Peenemuende, for all their sophisticated ideas about future space flights, had never thought of building anything comparable.[8]

The recommendations to the Bureau of Aeronautics quickly led to exploratory contracts with the Jet Propulsion Laboratory of the California Institute of Technology and the Aerojet General Corporation, a California firm with wartime experience in producing rocket fuels. Cal Tech's report, prepared by Homer J. Stewart and several associates and submitted in December 1945, verified the committee's calculations on the interrelationships of the orbit, the rocket's motor and fuel performance, the vehicle's structural characteristics, and payload. Aerojet's confirmation of the committee computations of the power obtainable from liquid hydrogen and liquid oxygen soon followed. Thus encouraged, BuAer assigned contracts to North American Aviation, Incorporated, and the Glenn L. Martin Company for preliminary structural design of the "ESV," the earth satellite vehicle, and undertook study of solar-powered devices to recharge the satellite's batteries and so lengthen their life. But as estimates put the cost of carrying the program beyond the preliminary stages at well over $5 million, a sum unlikely to be approved by the Navy high brass, ESV proponents sought Army Air Forces collaboration.[9] Curiously enough, with the compartmentation often characteristic of the armed services, BuAer apparently did not attempt to link its plans to those of the Naval Research Laboratory.[10]

In March 1946, shortly after NRL scientists had decided that a satellite was too difficult a project to attempt as yet, representatives of BuAer and the Army Air Forces agreed that "the general advantages to be derived from pursuing the satellite development appear to be sufficient to justify a major program, in spite of the fact that the obvious military, or purely naval applications in themselves, may not appear at this time to warrant the expenditure." General Curtis E. LeMay of the Air Staff did not concur. Certainly he was unwilling to endorse a joint Navy-Army program. On the contrary, Commander Hall noted that the general was resentful of Navy invasion into a field "which so obviously, he maintained, was the province of the AAF." Instead, in May 1946, the Army Air Forces presented its own proposition in the form of a feasibility study by Project Rand, a unit of the Douglas Aircraft Company and a forerunner of the RAND Corporation of California.[11] Like the scientists of the Bureau of Aeronautics committee, Project Rand mathematicians and engineers declared technology already equal to the task of launching a spaceship. The ship could be circling the earth, they averred, within five years, namely by mid-1951. They admitted that it could not be used as a carrier for an atomic bomb and would have no direct function as a weapon, but they stressed the advantages that would nevertheless accrue from putting an artificial satellite into orbit: "To visualize the impact on the world, one can imagine the consternation and admiration that would be felt here if the United States were to discover suddenly that some other nation had already put up a successful satellite."[12]

Officials at the Pentagon were unimpressed. Theodore von Kármán, chief mentor of the Army Air Forces and principal author of the report that became the research and development bible of the service, advocated research in the upper atmosphere but was silent about the use of an artificial satellite. Nor did Vannevar Bush have faith in such a venture. The most influential scientist in America of his day and in 1946 chairman of the Joint Army and Navy Research and Development Board, Bush was even skeptical about the possibility of developing within the foreseeable future the engineering skills necessary to build intercontinental guided missiles. His doubts, coupled with von Kármán's disregard of satellite schemes, inevitably dashed cold water on the proposals and helped account for the lukewarm reception long accorded them.[13]

Still the veto of a combined Navy-Army Air Forces program did not kill the hopes of advocates of a "space ship." While the Navy and its contractors continued the development of a scale model 3,000-pound-thrust motor powered by liquid hydrogen and liquid oxygen, Project Rand completed a second study for the Army Air Forces. But after mid-1947, when the Air Force became a separate service within the newly created Department of Defense, reorganization preoccupied its officers for a year or more, and many of them, academic scientists believed, shared General LeMay's indifference to research not immediately applicable to defense problems. At BuAer, on the other hand, a number of men continued to press for money to translate satellite studies into actual experiments. Unhappily for them, a Technical Evaluation Group of civilian scientists serving on the Guided Missiles Committee of the Defense Department's Research and Development Board declared in March 1948 that "neither the Navy nor the USAF has as yet established either a military or a scientific utility commensurate with the presently expected cost."[14] In vain, Louis Ridenour of Project Rand explained, as Hall had emphasized in 1945 and 1946, that

8

"the development of a satellite will be directly applicable to the development of an intercontinental rocket missile," since the initial velocity required for launching the latter would be "4.4 miles per second, while a satellite requires 5.4." [15]

In the hope of salvaging something from the discard, the Navy at this point shifted its approach. Backed up by a detailed engineering design prepared under contract by the Glenn L. Martin Company, BuAer proposed to build a sounding rocket able to rise to a record altitude of more than four hundred miles, since a powerful high-altitude test vehicle, HATV, might serve the dual purpose of providing hitherto unobtainable scientific data from the extreme upper atmosphere and at the same time dramatize the efficiency of the hydrogen propulsion system. Thus it might rally financial support for the ESV. But when *The First Annual Report of the Secretary of Defense* appeared in December 1948, a brief paragraph stating that each of the three services was carrying on studies and component designs for "the Earth Satellite Vehicle Program" evoked a public outcry at such a wasteful squandering of taxpayers' money; one outraged letter-writer declared the program an unholy defiance of God's will for mankind. That sort of response did not encourage a loosening of the military purse-strings for space exploration. Paper studies, yes; hardware, no. The Navy felt obligated to drop HATV development at a stage which, according to later testimony, was several years ahead of Soviet designs in its proposed propulsion system and structural engineering. [16]

In seeking an engine for an intermediate range ballistic missile, the Army Ordnance Corps, however, was able to profit from North American Aviation's experience with HATV design; an Air Force contract for the Navaho missile ultimately produced the engine that powered the Army's Jupiter C, the launcher for the first successful American satellite. Thus money denied the Navy for scientific research was made available to the Army for a military rocket. [17] Early in 1949 the Air Force requested the RAND Corporation, the recently organized successor to Project Rand, to prepare further utility studies. The paper submitted in 1951 concentrated upon analyzing the value of a satellite as an "instrument of political strategy," and again offered a cogent argument for supporting a project that could have such important psychological effects on world opinion as an American earth satellite. [18] Not until October 1957 would most of the officials who had read the text recognize the validity of that point.

In the meantime, research on the upper atmosphere had continued to nose forward slowly at White Sands and at the Naval Research Laboratory in Washington despite the transfer of some twenty "first line people" from NRL's Rocket Sonde Research Section to a nuclear weapons crash program. While the Navy team at White Sands carried on probes with the Aerobee,

HIGH-ALTITUDE TEST VEHICLE

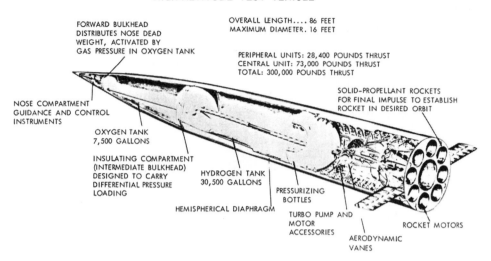

FORWARD BULKHEAD
DISTRIBUTES NOSE DEAD
WEIGHT, ACTIVATED BY
GAS PRESSURE IN OXYGEN TANK

OVERALL LENGTH.... 86 FEET
MAXIMUM DIAMETER. 16 FEET

PERIPHERAL UNITS: 28,400 POUNDS THRUST
CENTRAL UNIT: 73,000 POUNDS THRUST
TOTAL: 300,000 POUNDS THRUST

SOLID-PROPELLANT ROCKETS
FOR FINAL IMPULSE TO ESTABLISH
ROCKET IN DESIRED ORBIT

NOSE COMPARTMENT
GUIDANCE AND CONTROL
INSTRUMENTS

OXYGEN TANK
7,500 GALLONS

INSULATING COMPARTMENT
(INTERMEDIATE BULKHEAD)
DESIGNED TO CARRY
DIFFERENTIAL PRESSURE
LOADING

HYDROGEN TANK
30,500 GALLONS

PRESSURIZING
BOTTLES

HEMISPHERICAL DIAPHRAGM

TURBO PUMP AND
MOTOR
ACCESSORIES

AERODYNAMIC
VANES

ROCKET MOTORS

*The Navy's High-Altitude Test Vehicle (HATV). It was proposed in 1946
and was to have launched a satellite by 1951.*

by then known as "the workhorse of high altitude research," [19] a Bumper-
Wac under Army aegis—a V–2 with a Wac-Corporal rocket attached as a
second stage—made a record-breaking flight to an altitude of 250 miles in
February 1949. Shortly afterward tests began on the new sounding rocket
built for NRL by the Glenn L. Martin Company. Named "Neptune" at
first and then renamed "Viking," the first model embodied several important innovations: a gimbaled motor for steering, aluminum as the principal structural material, and intermittent gas jets for stabilizing the vehicle
after the main power cut off. Reaction Motors Incorporated supplied the
engine, one of the first three large liquid-propelled rocket power plants
produced in the United States. Viking No. 1, fired in the spring of 1949,
attained a 50-mile altitude; Viking No. 4, launched from shipboard in May
1950, reached 104 miles. Modest compared to the power displayed by the
Bumper-Wac, the thrust of the relatively small single-stage Viking nevertheless was noteworthy.[20]

While modifications to each Viking in turn brought improved performance, the Electron Optics Branch at NRL was working out a method of
using ion chambers and photon counters for x-ray and ultraviolet wavelengths, equipment which would later supply answers to questions about
the nuclear composition of solar radiation. Equally valuable was the
development of an electronic tracking device known as a "Single-Axis
Phase-Comparison Angle-Tracking Unit," the antecedent of "Minitrack,"
which would permit continuous tracking of a small instrumented body in

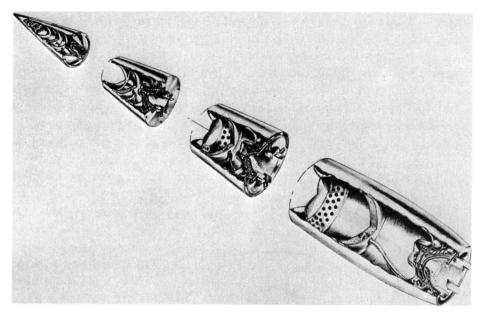

RAND Corporation proposal for a rocket to launch an "Earth Circling Satellite," 1951.

space. When the next to last Viking, No. 11, rose to an altitude of 158 miles in May 1954, the radio telemetering system transmitted data on cosmic ray emissions, just as the Viking 10, fired about two weeks before, had furnished scientists with the first measurement of positive ion composition at an altitude of 136 miles.[21] This remarkable series of successes achieved in five years at a total cost of less than $6 million encouraged NRL in 1955 to believe that, with a more powerful engine and the addition of upper stages, here was a vehicle capable of launching an earth satellite.

Essential though this work was to subsequent programs, the Naval Research Laboratory in the late 1940s and the 1950s was hampered by not having what John P. Hagen called "stable funding" for its projects. Hagen, head of the Atmosphere and Astrophysics Division, found the budgetary system singularly unsatisfactory. NRL had been founded in 1923, but a post-World-War-II reorganization within the Navy had brought the Office of Naval Research into being and given it administrative control of the Laboratory's finances. ONR allotted the Laboratory a modest fixed sum annually, but other Navy bureaus and federal agencies frequently engaged the Laboratory's talents and paid for particular jobs. The arrangement resembled that of a man who receives a small retainer from his employer but depends for most of his livelihood on fees paid him by his own clientele for special services. NRL's every contract, whether for design studies or hardware, had to be negotiated and administered either by ONR or by one of the permanent Navy bureaus—in atmospheric research, it was by the Navy Bureau of Aeronautics. The cancellation of a contract could seriously disrupt NRL functioning, as the years 1950 to 1954 illustrated.[22]

With the outbreak of the Korean War, the tempo of missile research heightened in the Defense Department. While the Navy was working on a guided missile launchable from shipboard and a group at NRL on radio interferometers for tracking it, rocketeers at Redstone Arsenal in Alabama were engaged in getting the "bugs" out of a North American Aviation engine for a ballistic missile with a 200-mile range, and RAND was carrying on secret studies of a military reconnaissance satellite for the Air Force. In June 1952 NRL got approval for the construction of four additional Vikings similar to Viking No. 10 to use in ballistic missile research, but eleven months later BuAer withdrew its support and canceled the development contract for a high-performance oxygen-ammonia engine that was to have replaced the less powerful Viking engine; this cancellation postponed by over three years the availability of a suitable power plant for the first stage of the future Vanguard rocket. Similarly in 1954 lack of funds curtailed an NRL program to design and develop a new liquid-propelled Aerobee-Hi probing rocket. At the request of the Western Development Division of the Air Force in July 1954, the Laboratory investigated the possible use of an improved Viking as a test vehicle for intercontinental ballistic missiles, ICBMs. The study, involving a solution of the "reentry problem," that is, how to enable a missile's warhead to return into the atmosphere without disintegrating before reaching its target, produced the design of an M–10 and M–15 Viking, the designations referring to the speeds, measured by Mach number, at which each would reenter the atmosphere. But the Air Force later let the development contracts to private industry.[23] In these years the Department of Defense was unwilling to spend more than token sums on research that appeared to have only remote connection with fighting equipment.

The creation of the National Science Foundation in May 1950 tended to justify that position, for one of the new agency's main functions was to encourage and provide support for basic research chiefly by means of grants-in-aid to American universities. The mission of the Army, Navy, and Air Force was national defense, that of the Foundation the fostering of scientific discovery. It was a responsibility of the Foundation to decide what lines of fundamental research most merited public financial aid in their own right, whereas other federal agencies must by law limit their basic research to fields closely related to their practical missions. While the Foundation's charter forbade it to make grants for applied research and development— the very area in which the military would often have welcomed financial

Aerobee-Hi.

Viking 10 on the launch pad at White Sands.

Bumper-Wac.

assistance—any government department could ask the National Academy of Sciences for help on scientific problems. The Academy, founded in 1863 as a self-perpetuating body advisory to but independent of the government, included distinguished men in every scientific field. When its executive unit, the National Research Council, agreed to sponsor studies for federal agencies, the studies sometimes involved more applied than pure research. The Academy's Research Council, and the Science Foundation, however, frequently worked closely together in choosing the problems to investigate.[24]

Certainly the composition of the ionosphere, the region that begins about fifty miles above the earth's surface, and the nature of outer space were less matters for the Pentagon than for the National Academy, the Science Foundation, and the academic scientific world. Indeed, the panel of volunteers which analyzed the findings from each instrumented V–2 shot and later appraised the results of Aerobee, Viking, and Aerobee-Hi flights contained from the first some future members of the Academy. Among the participants over the years were Homer J. Stewart and William H. Pickering of Cal Tech's Jet Propulsion Laboratory, Milton W. Rosen, Homer E. Newell, Jr., and John W. Townsend, Jr., of NRL, and James A. Van Allen of the Applied Physics Laboratory of the Johns Hopkins University and later a professor at the State University of Iowa. Under Van Allen's chairmanship, the Panel on Upper Atmosphere Rocket Research came to be a strong link between university physicists and the Department of Defense, a more direct link in several respects than that afforded by civilian scientists who served on advisory committees of the DoD's Research and Development Board.[25]

While the armed services were perforce confining their research and development programs chiefly to military objectives, no service wanted to discourage discussions of future possibilities. In the autumn of 1951 several doctors in the Air Force and a group of physicists brought together by Joseph Kaplan of the University of California, Los Angeles, met in San Antonio, Texas, for a symposium on the Physics and Medicine of the Upper Atmosphere. The participants summarized existing knowledge of the region named the "aeropause," where manned flight was not yet possible, and examined the problems of man's penetrating into that still unexplored area. The papers published in book form a year later were directly instrumental, Kaplan believed, in arousing enthusiasm for intensive studies of the ionosphere.[26]

A few months before the San Antonio sessions, the Hayden Planetarium of New York held a first annual symposium on space exploration, and about the same time the American Rocket Society set up an ad hoc Committee on Space Flight to look for other ways of awakening public interest and winning government support for interplanetary exploration. From a few dozen men who had followed rocket development in the early 1930s the society had grown to about two thousand members, some of them connected with the aircraft industry, some of them in government service, and some who were purely enthusiasts caught up by the imaginative possibilities of reaching out into the unknown. The committee met at intervals during the next two years at the Society's New York headquarters or at the Washington office of Andrew Haley, the Society's legal counsel, but not until Richard W. Porter of the General Electric Company sought out Alan T. Waterman, Director of the National Science Foundation, and obtained from him an assurance that the Foundation would consider a proposal, did a formal detailed statement of the committee's credo appear. Milton Rosen, the committee chairman and one of the principal engineers directing the development and tests of the Viking sounding rocket, then conceived and wrote the report advocating a thorough study of the benefits that might derive from launching an earth satellite. Completed on 27 November 1954, the document went to the Foundation early the next year.[27]

Without attempting to describe the type of launching vehicle that would be needed, the paper spelled out the reasons why space exploration would bring rich rewards. Six appendixes, each written by a scientist dealing with his own special field, pointed to existing gaps in knowledge which an instrumented satellite might fill. Ira S. Bowen, director of the Palomar Observatory at Mt. Wilson, explained how the clearer visibility and longer exposure possible in photoelectronic scanning of heavenly phenomena from a body two hundred miles above the earth would assist astronomers. Howard Schaeffer of the Naval School of Aviation Medicine wrote of the benefits of obtaining observations on the effects of the radiation from outer space upon living cells. In communications, John R. Pierce, whose proposal of 1952 gave birth to Telstar a decade later,[28] discussed the utility of a relay for radio and television broadcasts. Data obtainable in the realm of geodesy, according to Major John O'Keefe of the Army Map Service, would throw light on the size and shape of the earth and the intensity of its gravitational fields, information which would be invaluable to navigators and mapmakers. The meteorologist Eugene Bollay of North American Weather Consultants spoke of the predictable gains in accuracy of weather forecasting. Perhaps most illuminating to the nonscientifically trained reader was Homer E. Newell's analysis of the unknowns of the ionosphere which data accumulated over a period of days could clarify.

Confusing and complex happenings in the atmosphere, wrote Newell, were "a manifestation of an influx of energy from outer space." What was the nature and magnitude of that energy? Much of the incoming energy was absorbed in the atmosphere at high altitudes. From data transmitted from a space satellite five hundred miles above the earth, the earth-bound scientist might gauge the nature and intensity of the radiation emanating from the sun, the primary producer of that energy. Cosmic rays, meteors, and micrometeors also brought in energy. Although they probably had little effect on the upper atmosphere, cosmic rays, with their extremely high energies, produced ionization in the lower atmosphere. Low-energy particles from the sun were thought to cause the aurora and to play a significant part in the formation of the ionosphere. Sounding rockets permitted little more than momentary measurements of the various radiations at various heights, but with a satellite circling the earth in a geomagnetic meridian plane it should be possible to study in detail the low-energy end of the cosmic ray spectrum, a region inaccessible to direct observation within the atmosphere and best studied above the geomagnetic poles. Batteries charged by the sun should be able to supply power to relay information for weeks or months.

Contrary to what an indifferent public might have expected from rocket "crackpots," the document noted that "to create a satellite merely for the purpose of saying it has been done would not justify the cost. Rather, the satellite should serve useful purposes—purposes which can command the respect of the officials who sponsor it, the scientists and engineers who produce it, and the community who pays for it." The appeal was primarily to the scientific community, but the intelligent layman could comprehend it, and its publication in an engineering journal in February 1955 gave the report a diversified audience.[29]

A number of men in and outside government service meantime had continued to pursue the satellite idea. In February 1952 Aristid V. Grosse of Temple University, a key figure in the Manhattan Project in its early days, had persuaded President Truman to approve a study of the utility of a satellite in the form of an inflatable balloon visible to the naked eye from the surface of the earth. Aware that Wernher von Braun, one of the German-born experts from Peenemuende, was interested, the physicist took counsel with him and his associates at Redstone Arsenal in Huntsville, Alabama. Fifteen months later Grosse submitted to the Secretary of the Air Force a description of the "American Star" that could rise in the West. Presumably because the proposed satellite would be merely a show piece without other utility, nothing more was heard of it.[30]

A series of articles in three issues of *Collier's*, however, commanded wide attention during 1952. Stirred by an account of the San Antonio symposium as Kaplan described it over the lunch table, the editors of the magazine engaged Wernher von Braun to write the principal pieces and obtained shorter contributions from Kaplan, Fred L. Whipple, chairman of the Harvard University Department of Astronomy, Heinz Haber of the Air Force Space Medicine Division, the journalist Willy Ley, and others. The editors' comment ran: "What are we waiting for?", an expression of alarm lest a communist nation preempt outer space before the United States acted and thereby control the earth from manned space platforms equipped with atomic bombs. On the other hand, von Braun's articles chiefly stressed the exciting discoveries possible within twenty-five years if America at once began building "cargo rockets" and a wheel-shaped earth-circling space station from which American rocket ships could depart to other planets and return. Perhaps because of severe editing to adapt the material to popular consumption, the text contained little or no technical data on how these wonders were to be accomplished; the term "telemetry" nowhere appeared. But the articles, replete with illustrations in color, and a subsequent Walt Disney film fanned public interest and led to an exchange of letters between von Braun and S. Fred Singer, a brilliant young physicist at the University of Maryland.[31]

At the fourth Congress of the International Astronautics Federation in Zurich, Switzerland, in summer 1953, Singer proposed a Minimum Orbital Unmanned Satellite of the Earth, MOUSE, based upon a study prepared two years earlier by members of the British Interplanetary Society who had predicated their scheme on the use of a V–2 rocket. The Upper Atmosphere Rocket Research Panel at White Sands in turn discussed the plan in April 1954, and in May Singer again presented his MOUSE proposal at the Hayden Planetarium's fourth Space Travel Symposium. On that occasion Harry Wexler of the United States Weather Bureau gave a lecture entitled, "Observing the Weather from a Satellite Vehicle." [32] The American public was thus being exposed to the concept of an artificial satellite as something more than science fiction.

By then, Commander George Hoover and Alexander Satin of the Air Branch of the Office of Naval Research had come to the conclusion that recent technological advances in rocketry had so improved the art that the feasibility of launching a satellite was no longer in serious doubt. Hoover therefore put out feelers to specialists of the Army Ballistic Missile Agency at Huntsville. There von Braun, having temporarily discarded his space platform as impractical, was giving thought to using the Redstone rocket to place a small satellite in orbit. Redstone, a direct descendant of the V–2, was, as one man described it, a huge piece of "boiler plate," sixty-nine feet long, seventy inches in diameter, and weighing 61,000 pounds, its power plant using liquid oxygen as oxidizer and an alcohol-water mixture as fuel. A new Redstone engine built by the Rocketdyne Division of North American Aviation, Inc., and tested in 1953 was thirty percent lighter and thirty-four percent more powerful than that of the V–2.[33] If Commander Hoover knew of the futile efforts of BuAer in 1947 to get Army Air Forces collaboration on a not wholly dissimilar space program, that earlier disappointment failed to discourage him. And as he had reason to believe he could now get Navy funds for a satellite project, he had no difficulty in enlisting von Braun's interest. At a meeting in Washington arranged by Frederick C. Durant, III, past president of the American Rocket Society, Hoover, Satin, von Braun, and David Young from Huntsville discussed possibilities with Durant, Singer, and Fred Whipple, the foremost American authority on tracking heavenly bodies. The consensus of the conferees ran that a slightly modified Redstone rocket with clusters of thirty-one Loki solid-propellant rockets for upper stages could put a five-pound satellite into orbit at a minimum altitude of 200 miles. Were that successful, a larger satellite equipped with instruments could follow soon afterward. Whipple's judgment that optical tracking would suffice to trace so small a satellite at a distance of 200 miles led the group to conclude that radio tracking would be needless.[34]

Whipple then approached the National Science Foundation begging it to finance a conference on the technical gains to be expected from a satellite and from "the instrumentation that should be designed well in advance of the advent of an active satellite vehicle." The Foundation, he noted some months later, was favorable to the idea but in 1954 took no action upon it.[35] Commander Hoover fared better. He took the proposal to Admiral Frederick R. Furth of the Office of Naval Research and with the admiral's approval then discussed the division of labor with General H. T. Toftoy and von Braun at Redstone Arsenal. The upshot was an agreement that the Army should design and construct the booster system, the Navy take responsibility for the satellite, tracking facilities, and the acquisition and analysis of data. No one at ONR had consulted the Naval Research Laboratory about the plan. In November 1954 a full description of the newly named Project Orbiter was sent for critical examination and comment to Emmanuel R. Piore, chief scientist of ONR, and to the government-owned Jet Propulsion Laboratory in Pasadena which handled much of the Army Ballistic Missile Agency's research. Before the end of the year, the Office of Naval Research had let three contracts totaling $60,000 for feasibility analyses or design of components for subsystems. Called a "no-cost satellite," Orbiter was to be built largely from existing hardware.[36]

At this point it is necessary to examine the course scientific thought had been taking among physicists of the National Academy and American universities, for in the long run it was their recommendations that would most immediately affect governmental decisions about a satellite program. This phase of the story opens in spring 1950, at an informal gathering at James Van Allen's home in Silver Spring, Maryland. The group invited by Van Allen to meet with the eminent British geophysicist Sydney Chapman consisted of Lloyd Berkner, head of the new Brookhaven National Laboratory on Long Island, S. Fred Singer, J. Wallace Joyce, a geophysicist with the

*A Redstone rocket on the static-firing test stand at the Army
Ballistic Missile Agency, Huntsville, Alabama.*

Navy BuAer and adviser to the Department of State, and Ernest H. Vestine
of the Department of Terrestrial Magnetism of the Carnegie Institution.
As they talked of how to obtain simultaneous measurements and observations of the earth and the upper atmosphere from a distance above the
earth, Berkner suggested that perhaps staging another International Polar
Year would be the best way. His companions immediately responded enthu-
siastically. Berkner and Chapman then developed the idea further and put
it into form to present to the International Council of Scientific Unions.
The first International Polar Year had established the precedent of international scientific cooperation in 1882 when scientists of a score of nations

Meeting on Project Orbiter, 17 March 1955 in Washington, D. C.: left to right, seated, Commander George W. Hoover, Office of Naval Research; Frederick C. Durant III, Arthur D. Little, Inc.; James B. Kendrick, Aerophysics Development Corp.; William A. Giardini, Alabama Tool and Die; Philippe W. Newton, Department of Defense; Rudolf H. Schlidt, Army Ballistic Missile Agency (ABMA); Gerhard Heller, ABMA; Wernher von Braun, ABMA.   Standing—Lieutenant Commander William E. Dowdell, USN; Alexander Satin, ONR; Commander Robert C. Truax, USN; Liston Tatum, International Business Machines (IBM); Austin W. Stanton, Varo, Inc.; Fred L. Whipple, Harvard Observatory; George W. Petri, IBM; Lowell O. Anderson, Office of Naval Research; Milton W. Rosen, Naval Research Laboratory. (Smithsonian Institution photo.)

agreed to pool their efforts for a year in studying polar conditions. A second International Polar Year took place in 1932. Berkner's proposal to shorten the interval to 25 years was timely because 1957–1958, astronomers knew, would be a period of maximum solar activity.[37] European scientists subscribed to the plan. In 1952 the International Council of Scientific Unions appointed a committee to make arrangements, extended the scope of the study to the whole earth, not just the polar regions, fixed the duration at eighteen months, and then renamed the undertaking the International Geophysical Year, shortened in popular speech to IGY. It eventually embraced sixty-seven nations.[38]

In the International Council of Scientific Unions the National Academy of Sciences had always been the adhering body for the United States. The Council itself, generally called ICSU, was and is the headquarters unit of a nongovernmental international association of scientific groups such as the International Union of Geodesy and Geophysics, the International Union of Pure and Applied Physics, the International Scientific Radio Union, and others. When plans were afoot for international scientific programs which needed governmental support, Americans of the National Academy naturally looked to the National Science Foundation for federal funds. Relations between the two organizations had always been cordial, the Foundation often turning for advice to the Academy and its secretariat, the National Research Council, and the Academy frequently seeking financing for projects from the Foundation. At the end of 1952 the Academy appointed a United States National Committee for the IGY headed by Joseph Kaplan to plan for American participation. The choice of Kaplan as chairman strengthened the position of men interested in the upper atmosphere and outer space.

During the spring of 1953 the United States National Committee drafted a statement which the International Council later adopted, listing the fields of inquiry which IGY programs should encompass—oceanographic phenomena, polar geography, and seismology, for example, and, in the celestial area, such matters as solar activity, sources of ionizing radiations, cosmic rays, and their effects upon the atmosphere.[39] In the course of the year the Science Foundation granted $27,000 to the IGY committee for planning, but in December, when Hugh Odishaw left his post as assistant to the director of the Bureau of Standards to become secretary of the National Committee, it was still uncertain how much further support the government would give IGY programs. Foundation resources were limited. Although in August Congress had removed the $15,000,000 ceiling which the original act had placed on the Foundation's annual budget, the appropriation voted for FY 1954 had totaled only $8 million. In view of the Foundation's other commitments, that sum seemed unlikely to allow for extensive participation in the IGY. In January 1954 the National Committee asked for a total of $13 million. Scientists' hopes rose in March when President Eisenhower announced that, in contrast to the $100 million spent in 1940 on federal support of research and development, he was submitting a $2-billion research and development budget to Congress for FY 1955. Hope turned to gratification in June when Congress authorized for the IGY an over-all expenditure of $13 million as requested and in August voted for FY 1955 an appropriation of $2 million to the National Science Foundation for IGY preparations.[40]

Thus reassured, the representatives from the National Academy set out in the late summer for Europe and the sessions of the International Scientific Radio Union, known as URSI, and the International Union of Geodesy and Geophysics, IUGG. As yet none of the nations pledged to take part in the IGY had committed itself to definite projects. The U.S.S.R. had not yet joined at all, although Russian delegates attended the meetings. Before the meetings opened, Lloyd V. Berkner, president of the Radio Union and vice president of Comité Spéciale de l'Année Géophysique Internationale (CSAGI) set up two small informal committees under the chairmanship of Fred Singer and Homer E. Newell, Jr., respectively, to consider the scientific utility of a satellite. The National Academy's earlier listing of IGY objectives had named problems requiring exploration but had not suggested specific means of solving them. For years physicists and geodesists had talked wistfully of observing the earth and its celestial environment from above the atmosphere. Now, Berkner concluded, was the time to examine the possibility of acting upon the idea. Singer was an enthusiast who inclined to brush aside technical obstacles. Having presented MOUSE the preceding year and shared in planning Project Orbiter, he was a persuasive proponent of an IGY satellite program. Newell of NRL was more conservative, but he too stressed to IUGG the benefits to be expected from a successful launching of an instrumented "bird," the theme that he incorporated in his later essay for the American Rocket Society. URSI and IUGG both passed resolutions favoring the scheme. But CSAGI still had to approve. And there were potential difficulties.

Hence on the eve of the CSAGI meeting in Rome, Berkner invited ten of his associates to his room at the Hotel Majestic to review the pros and cons, to make sure, as one man put it, that the proposal to CSAGI was not just a "pious resolution" such as Newton could have submitted to the Royal Society. The group included Joseph Kaplan, U.S. National Committee chairman, Hugh Odishaw, committee secretary, Athelstan Spilhaus, Dean of the University of Minnesota's Institute of Technology, Alan H. Shapley of the National Bureau of Standards, Harry Wexler of the Weather Bureau, Wallace Joyce, Newell, and Singer. The session lasted far into the night. Singer outlined the scientific and technical problems—the determination of orbits, the effects of launching errors, the probable life of the satellite, telemetering and satellite orientation, receiving stations, power supplies, and geophysical and astrophysical applications of data. Newell, better versed than some of the others in the technical difficulties to be overcome, pointed out that satellite batteries might bubble in the weightless environment of space, whereupon Spilhaus banged his fist and shouted: "Then we'll get batteries that won't!" Singer's presentation was exciting, but the question remained whether an artificial body of the limited size and weight a rocket could as yet put into orbit could carry enough reliable instrumentation to prove of sufficient scientific value to warrant the cost; money and effort poured into that project would not be available for other research, and to attempt to build a big satellite might be to invite defeat.

Both Berkner and Spilhaus spoke of the political and psychological prestige that would accrue to the nation that first launched a man-made satellite. As everyone present knew, A. N. Nesmeyanov of the Soviet Academy of Sciences had said in November 1953 that satellite launchings and moon shots were already feasible; and with Tsiolkovskiy's work now recognized by Western physicists, the Americans had reason to believe in Russian scientific and technological capabilities. In March 1954 Moscow Radio had exhorted Soviet youth to prepare for space exploration, and in April the Moscow Air Club had announced that studies in interplanetary flight were beginning. Very recently the U.S.S.R. had committed itself to IGY participation. While the American scientists in September 1954 did not discount the possible Russian challenge, some of them insisted that a satellite experiment must not assume such emphasis as to cripple or halt upper atmosphere research by means of sounding rockets. The latter was an established useful technique that could provide, as a satellite in orbit could not, measurements at a succession of altitudes in and above the upper atmosphere, measurements along the vertical instead of the horizontal plane. Nevertheless at the end of the six-hour session, the group unanimously agreed to urge CSAGI to endorse an IGY satellite project.[41]

During the CSAGI meeting that followed, the Soviet representatives listened to the discussion but neither objected, volunteered comment, nor asked questions. On 4 October CSAGI adopted the American proposal: "In view," stated that body,

> of the great importance of observations during extended periods of time of extra-terrestrial radiations and geophysical phenomena in the upper atmosphere, and in view of the advanced state of present rocket techniques, CSAGI recommends that thought be given to the launching of small satellite vehicles, to their scientific instrumentation, and to the new problems associated with satellite experiments, such as power supply, telemetering, and orientation of the vehicle.[42]

What had long seemed to most of the American public as pure Jules Verne and Buck Rogers fantasy now had the formal backing of the world's most eminent scientists.

Thus by the time the United States Committee for the IGY appointed a Feasibility Panel on Upper Atmosphere Research, three separate, albeit interrelated, groups of Americans were concerned with a possible earth satellite project: physicists, geodesists, and astronomers intent on basic research; officers of the three armed services looking for scientific means to military ends; and industrial engineers, including members of the American Rocket Society, who were eager to see an expanding role for their companies. The three were by no means mutually exclusive. The dedicated scientist, for instance, in keeping with Theodore von Kármán's example as a founder and official of the Aerojet General Corporation, might also be a shareholder in a research-orientated electronics or aircraft company, just as the industrialist might have a passionate interest in pure as well as applied science, and the military man might share the intellectual and practical interests of both the others. Certainly all three wanted improvements in equipment for national defense. Still the primary objective of each group differed from those of the other two. These differences were to have subtle effects on Vanguard's development. Although to some people the role of the National Academy appeared to be that of a Johnny-come-lately, the impelling force behind the satellite project nevertheless was the scientist speaking through governmental and quasi-governmental bodies.

2
SEEKING GOVERNMENT SUPPORT FOR A SATELLITE PROGRAM

NINE and a half years after studies of an artificial satellite had begun in the United States, top level government officials gave the idea serious consideration. Indeed until then few of them had had more than fleeting exposure to the seemingly extravagant notion of creating a man-made moon. Due, however, to interest in plans for the IGY, matters came to a head in the first half of 1955. In January, Radio Moscow announced that a satellite launching might be expected in the not distant future. In Washington, while the National Science Foundation was examining the American Rocket Society's plea and copies of the Orbiter proposal were going the rounds in the Defense Department, the National Academy's IGY Committee, having spent the autumn in sounding out American scientific opinion, set up a Technical Panel on Rocketry, consisting of Kaplan, Odishaw, Newell, Singer, Spilhaus, Whipple, Van Allen, Nathaniel Gerson, Bernhard Haurwitz of the Academy's Meteorology Panel, and Gerhardt F. Schilling of the IGY staff. At the first meeting in late January 1955 the panel created a special study group called the Subcommittee on the Technical Feasibility of a Long Playing Rocket, that is, a satellite. The name coined by Joseph Kaplan not only was descriptive but also provided protective coloration, a safeguard against premature publicity about a plan the Academy might decide to reject.

Why further studies of feasibility seemed necessary to the panel may at first puzzle the layman, inasmuch as over the years feasibility reports had accumulated steadily at the Pentagon. But not more than five or six men at the Academy had ever seen the earlier Navy and RAND studies. Besides, who could be sure that what would suffice for a military satellite would also do for a scientific one? And, as Homer Newell observed, there was feasibility and feasibility. It depended upon the scale of the plan and the effort to be expended—factors related to costs and to expected rewards. The American delegates who thrashed over the problem in Rome had not seen entirely eye to eye about how much or how little was worth trying for and what would be attainable within the eighteen-month span of the IGY, July 1957 through December 1958. Hence the "LPR" subcommittee to appraise the details of the evidence.[1]

William H. Pickering, director of the Jet Propulsion Laboratory at Cal Tech, John W. Townsend, assistant head of NRL's Rocket Sonde Branch, and Milton Rosen of Viking fame, who composed the LPR sub-committee, were to report to the panel before 10 March 1955 on the feasibility and "geophysical possibilities" of an LPR, on the needed controls, the engine, the desired orbit, manpower requirements, and estimated costs. Three men better qualified for the job would have been hard to find. All three had served on the Upper Atmosphere Rocket Research Panel at White Sands and were familiar with the engineering problems involved in rocketry and with most, if not all, of the satellite plans prepared to date. Pickering had seen the Project Orbiter proposal at JPL, and Townsend and Rosen had studied it when the Office of Naval Research sent a copy for review to NRL's Atmosphere and Astrophysics Division and its Rocket Development Branch.

In the course of the investigation undertaken for the Air Force in 1954 on guided missile reentry into the atmosphere, Rosen, Townsend, and other NRL rocket specialists had gone a long way toward solving the problem of putting a satellite into orbit.[2] After examining the Army–ONR satellite plan, they concluded they could offer something better, a system which would obviate the weaknesses of Orbiter's low injection altitude, lack of guidance, the dubious reliability of the upper stages, and the dependence on optical tracking. Long after the event, critics would imply that NRL was guilty of reprehensible oneupmanship. The authors of the plan based on the Viking rocket believed, however, that the better should always supersede the inferior, and they had on hand the design studies made for the Air Force. Rosen, moreover, had discussed the NRL idea with the chief scientist of the Office of Naval Research who, while pointing out that ONR was at least partly committed to the Redstone launching scheme, saw the advantages of an alternative and encouraged Rosen and his associates to complete their counter proposal. The NRL scheme of a three-stage launcher, an instrumented satellite, and an electronic tracking system was taking form while the LPR subcommittee was drafting its recommendations for the Academy's rocketry panel.[3]

Members of the LPR subcommittee had made it clear to the panel from the beginning that they were not prepared to advocate any one satellite plan over any other. In a preliminary report in early February 1955 they declared that existing propulsion systems, if given somewhat more power, could lift a ten-pound payload to the necessary altitude, and existing control and guidance components, "after an appreciable amount of development work," could direct the bird into orbit. If enough competent men and enough money were assigned to the task, the feat could be accomplished within two or three years. On 9 March the panel and subcommittee prepared their findings to present to the United States National Committee the next day. Rosen, speaking for the subcommittee, pointed out that any one of three launching techniques would suffice for an LPR: that is, a large single-stage rocket, of which three were already available, could release "a number of small rockets at or near the top of the flight path," a method which would require guidance accuracy to within a one degree arc; or, second, a two- or three-stage launcher, though more difficult to guide accurately, could carry a large instrumented payload; or, third, around the most powerful engine then under development a new test vehicle could be built which would have the capability of putting into orbit a much bigger satellite with many more elaborate scientific measuring devices. The drawback in the last lay in the amount of time and study that would be needed to design, construct, and test the vehicle. Unbeknownst to some panel members, these three possibilities bore fairly close resemblance to Orbiter, to NRL's as yet uncompleted proposal, and to a system calling for use of the Air Force's only partly developed Atlas rocket. The Air Force at that moment was just beginning to solicit design studies for a military satellite.

The panel was concerned only with making sure that hope was not beclouding judgment on the feasibility of a satellite big enough and well enough equipped with instruments to transmit scientifically useful data during the IGY. Spilhaus, to be sure, arguing that "we must crawl before we walk," thought instrumentation unimportant in a first man-made satellite; study of the orbital pattern would supply scientists with ample data to begin with. His associates considered him over-cautious. Whipple remarked that a one-pound satellite would be valueless, too small to be observable from the ground; a ten-pound, on the contrary, about twenty inches in diameter painted white or with reflecting surfaces would be optically visible at twilight and dawn and trackable by binoculars and telescopic camera. In an equatorial orbit with a 250-mile perigee, its closest point to earth, and a 500-mile apogee, the most distant, a ten-pound instrumented body could relay by radio data which would enable scientists to make precise observations of the orbit, fix intercontinental distances to within one hundred feet, and determine the mass and density distribution of the earth's crust. Newell, concurring in Whipple's view that a solar or a nuclear power supply would be impractical for the next three years, described a recent NRL engineering study of instrumentation for a fifteen-inch, thirty- to fifty-pound satellite in equatorial orbit. The outcome of the panel's discussion was a unanimous endorsement of a satellite project and the conclusion that use of the second launching technique named and a thirty-pound instrumented orbiting body held the most promise of success. The estimate of costs was vague.[4]

The report reached the Executive Committee of the United States National Committee (USNC) on 10 March. What response it would evoke was uncertain, for a good many Academy officials and some at the Science Foundation had misgivings about the wisdom of including the project in the American IGY program. One reason was the risk involved in having to depend on a totally untried research technique; if, the judgment of the LPR group and the panel notwithstanding, the launching attempts failed to put a satellite in orbit, the United States would have invested a large sum of money only to win ridicule and taxpayers' censure. Any such outcome would also weaken congressional confidence in the National Science Foundation. Second was the likelihood that so spectacular a project, if adopted, would overshadow every other part of the IGY, thereby belittling undertakings of equal scientific importance, albeit, in Hugh Odishaw's words, with "less sex appeal." Third was the possibility that, in spite of the USNC's every effort, the project might take on a military aura that would conflict with IGY purposes.[5]

In accord with Academy policy, the IGY secretariat always kept itself out of the limelight, in a position of anonymity, but people in close touch with Academy affairs were aware that Hugh Odishaw played a large part in preventing the satellite proposal from dropping out of sight amid the flurry of planning for more orthodox IGY programs. He was constantly on the scene, as committee members were not, and he carried on most of the IGY correspondence. Trained both as a humanist and as an engineer, the Executive Secretary believed Academy policymakers must carefully examine the satellite scheme along with the rocket experts' recommendations. The moment had now come for the IGY Committee to support or bury the plan.

The official minutes of the Executive Committee meeting were cryptic. Containing no hint of any controversial discussion, they merely recorded the committee decision: to ask for ten thirty- to fifty-pound instrumented satellites in hopes of getting up five that could circle the earth for at least two weeks at an altitude of 250 miles in an equatorial orbit. That program would require five ground stations and the services of twenty-five scientists. The costs would probably run to over $7 million. Joseph Kaplan then dispatched letters to the Academy's president, Detlev Bronk, and to Alan T. Waterman, head of the Science Foundation, stating the reasons for requesting government support for such a project. Kaplan's letter opened with an explanation that a fifty-pound "bird" which international agencies could inspect before launching and track while in flight would be in accord with the CSAGI recommendation of 4 October 1954. Kaplan quoted that resolution in full. In concluding he wrote: "The Executive Committee of the U.S. National Committee, basing its opinion on the study of the expert panel on rocketry, feels that a small artificial satellite for geophysical purposes is feasible during the IGY if action is initiated promptly, and that realization of such a satellite would give promise of yielding results of geophysical interest." [6]

While counting on the United States Treasury to foot the bills for this difficult undertaking, the Executive Committee foresaw that the organizational ramifications of authority and responsibility always to be expected in government offices were likely to be more complex than usual. If intricate crisscrossing of bureaucratic channels later observable inspired academics to wish that a private organization, untrammeled by the checks and balances that attend government operations, had taken sole charge, that arrangement was from the first patently impossible. Costs alone would pose a nearly insuperable obstacle. Men attached to the Academy's National Committee would naturally determine what data were most wanted and would assign to qualified scientists the task of designing and making the satellite and the instruments necessary to obtain the desired information. But the launching vehicle inevitably would come within the purview of the Department of Defense and hence under DoD security regulations. Moreover, use of a military launching site, a virtual necessity, would mean government surveillance. Yet under government sponsorship the circuitous chain of command and the number of federal agencies that would be involved were bound to create delays, regardless of which agency or who took charge. In the spring of 1955 the mere mechanics of getting government acceptance were elaborate.

Since over five months had already elapsed since announcement of the CSAGI resolution, the Executive Committee chose to start negotiations without waiting for formal USNC endorsement. Detlev Bronk, as president of the Academy, and Alan Waterman, as director of the National Science Foundation, were to make the first overtures. Because the launching of an instrument-carrying earth satellite would be expensive, and because, as a contribution to the IGY, it would entail sharing information with other nations, approval of the plan had to come from the President of the United States. Before he made a decision, the proposal would have to undergo the scrutiny of the Science Foundation's National Science Board, and, as an enterprise that might affect foreign relations, would have to receive the blessings of the Department of State. Next the President would consult his Scientific Advisory Committee at the White House, his special assistants on Security Affairs and Economic Affairs, the directors of CIA and the Bureau of the Budget, and the National Security Council. Once convinced that the undertaking was worth the risks, the President would decide what agency should do what. Even if a special appropriation were unnecessary, probably Congress would later have to vote money to the Science Foundation for the project. Although the Executive Committee of the USNC was anxious to have matters settled before the beginning of the new fiscal year on 1 July, there was nothing to do but present the case, explain the whys and wherefores, and then wait. Optimists at the Academy hoped that a forceful presentation would persuade the President not only to support the venture but to give it the standing of a major national enterprise comparable to a Manhattan Project.

Waterman and Bronk lost no time in getting in touch with the key people. On 22 March the two men, accompanied by Lloyd Berkner, acquainted Robert Murphy of the Department of State with the project and asked for the department's approval. They then took the Academy's recommendation to the White House and awakened President Eisenhower's interest in it. A conference with Secretary of Defense Charles Wilson followed. Wilson, averse to all military excursions into basic research, was unenthusiastic but referred the proposition to his Assistant Secretary for Research and Development, Donald Quarles. Quarles already had before him a copy of the Orbiter proposal, a note from the Assistant Secretary of the Navy for Air commenting favorably upon it, and two memoranda of 3 March forwarded from the NRL Rocket Development Branch, one memo written by Milton Rosen describing the utility of an M–10 Viking as a satellite launcher and one entitled "Proposal for Minimum Trackable Satellite (Minitrack)" prepared by John T. Mengel and Roger Easton. At the same time, as Secretary Quarles knew, the Air Force had in progress plans for a military satellite using the Atlas or the Titan long-range ballistic missile. Confronted with three service schemes, Quarles secured from Secretary Wilson instructions to commit no funds to any of the three until the General Sciences Coordinating Committee had reviewed the situation. The completed NRL proposal combining the data contained in the two memos of early March was in the hands of the Coordinating Committee by mid-April.

Discussions with Alan Waterman, however, had strengthened Quarles' convictions that one plan or another was worth pursuing. When the Coordinating Committee, with representatives on it from all three services, recommended support for each of the three projects and a reappraisal at the end of six months, Quarles rejected the arrangement as wasteful. Instead he appointed an ad hoc Group on Special Capabilities composed of eight distinguished civilian scientists to assess the relative merits of the proposals, if and when the President decided to proceed. The group met once in early

May to map out a work schedule, but then waited for a green light from the White House.[7]

On 6 May the Science Foundation received from Kaplan of the USNC's Executive Committee its estimated budget of $9,734,500 for "(i) approximately ten 'birds' and five observation stations, including the necessary scientific instrumentation, related equipment, and minimum civilian scientific staff and . . . (ii) approximately ten vehicles and their associated flight instrumentation. Cost estimates for (i) are $2,234,500; for (ii) $7,500,000." The committee had come to believe that item (ii) ought to be part of the IGY budget in order to emphasize the nonmilitary nature of the program, to minimize classification problems, and to keep clear lines of demarcation between the National Committee and the Department of Defense. The $7.5 million for the ten vehicles was to include "procurement, construction, and necessary system design and development." The other $2,234,500 was to cover "procurement, construction and design relating to the 'birds' and observing equipment." The committee felt it ought to have the money by 1 July, inasmuch as the USNC had made its original recommendations contingent upon starting the program "promptly."

When the National Committee met on 18 May, Merle Tuve of the Carnegie Institution, to the consternation of some committee members, objected to the proposition: it would entangle the Academy in Defense security regulations, thereby barring the free exchange of data with other nations and defeating the very purposes of the IGY; the program was likely to net scientists too little information to justify the risk of unwholesome political repercussions throughout the world. Supported by Lloyd Berkner's eloquent defense of the plan, his associates overrode Tuve: they concluded that under Science Foundation and Academy sponsorship the program would remain civilian in nature and would prove rewarding. But Eisenhower, torn between his loyalty to Secretary Wilson and his profound respect for Waterman and Quarles, still hesitated.[8] How would Congress and American taxpayers respond to news that the United States proposed to spend millions of dollars to obtain information that it would then give free to other nations, Iron Curtain countries included?

In mid-May a military evaluation prepared by Quarles' staff reached the President's special assistant, Nelson A. Rockefeller, and, with Rockefeller's comments appended, went to the Secretary of the Treasury, the directors of CIA and the Bureau of the Budget, the Chairman of the Joint Chiefs of Staff, and then the National Security Council. Under the heading "General Considerations," the paper noted that "recent studies" within the Defense Department indicated that an adaptation of existing rocket components could launch a five- to ten-pound satellite before the end of 1958; a panel of the President's Scientific Advisory Committee had declared such a program warranted partly because of its scientific merit but especially because it would test "Freedom of Space" as a principle of international law. "On April 15, 1955," the exposition stated,

> the Soviet Government announced that a permanent high-level, interdepartmental commission for interplanetary communications had been created in the Astronomics Council of the U.S.S.R. Academy of Sciences. A group of Russia's top scientists is now believed to be working on a satellite program. In September 1954 the Soviet Academy announced the establishment of the Tsiolkovsky Gold Medal which would be awarded every three years for outstanding work in the field of interplanetary communications.

A scientific satellite should furnish data on air drag at extreme altitudes by means of observation and analysis of the orbital decay, and on "the shape of and gravitational field of the earth." Information obtainable on the ion content of the ionosphere, moreover, should benefit missile research and defense communications. An annex to the paper noted that a small satellite would not serve for military surveillance, but a successful orbiting would be a step toward that goal.[9]

Rockefeller's memorandum gave the proposal enthusiastic endorsement. "I am impressed," he wrote, "by the costly consequences of allowing the Russian initiative to outrun ours through an achievement that will symbolize scientific and technological advancement to people everywhere. The stake of prestige that is involved makes this a race that we cannot afford to lose." The more guarded military comment spoke of "considerable prestige and psychological benefits" for the nation that was first successful since a demonstration of such advanced technology and its "unmistakable relationship to intercontinental ballistic missile technology might have important repercussions on the political determination of free world countries to resist Communist threats." The military appraisal subtly conveyed the impression that the Soviets were unlikely to outstrip the United States in a satellite endeavor. Both statements, however, underscored the idea that here would be a race between the two nations. Post-Sputnik declarations that the United States was not racing with the U.S.S.R. obviously ignored the points of view expressed at the White House level in May 1955.[10] The United States, Rockefeller argued, should promptly announce to the world that it was embarking upon a scientific project the results of which would be made available to all nations. At the same time, to fend off any Russian attempt to label the satellite a threat to peace, the announcement must stress that military missions were not involved. Lest the Soviets claim to have already launched a satellite or to be working on one with a shorter timetable for launching, the American government must publicize its plans quickly. Concurrent with the development of a small, simple satellite, the United States should pursue work on a more sophisticated type so that the U.S.S.R. could not undercut American prestige by putting a bigger, more impressive body into orbit on the heels of the American.

The initial presentation declared that government support for a scientific satellite and recognition of its peaceful purposes must not prejudice freedom of action to develop military satellites. Nor should the project delay major defense programs. "The satellite itself and much information as to its orbit would be public information; the means of launching would be classified." Development of the vehicle would probably cost $10 to $15 million, the tracking equipment $2.5 million, and the logistics for launching and tracking another $2.5 million, all told $15 to $20 million. Quarles' staff, after examining the USNC Executive Committee's figures, had deliberately doubled them. As the size, complexity, and longevity of the satellite and the duration of the scientific observation program would affect costs, the staff paper noted that the total might well run higher than the estimate. The $15 to $20 million excluded the costs of research and development work that was already part of military programs. Orbiter and Viking both held promise, but exploratory studies should go forward on a backup program based upon the Air Force Atlas missile and the Aerobee research rocket.

A technical annex analyzed in somewhat greater detail the scientific and military value of the proposed project. While the amount of information a satellite might supply would depend on its size and "whether" it could carry instruments, precise observation of the orbital path of even a small, inert, body should give data on air density, pressure, and temperatures at high altitudes—information important for both manned aircraft and missiles— and at the same time furnish more exact knowledge about the shape of the earth. From an instrumented satellite accurate data should be forthcoming on the position of the continents, the gravitational constants over long distances, the earth's semimajor axis, and the rate of the earth's rotation. Organizing and operating any satellite launching should give missile crews useful experience. Research in electronic tracking would promote the development of antimissile missiles, since the satellite would have the speed and altitude of an ICBM. Optical tracking, though cheaper, would be possible only in clear weather and then only during a few minutes at dawn and a few at dusk.

Some explanation followed about the advantages of an orbital plane inclined about thirty-five degrees to the equator in contrast to a polar orbit. With a 200-mile perigee and a 1,000-mile apogee, the satellite would circle the earth in about ninety minutes. A polar orbit would require observation stations in the arctic regions, whereas launching at an inclination to the equator from the Air Force Missile Test Center at Cape Canaveral, Florida, would permit the use of tracking stations at the Navy centers at Point Mugu and Inyokern, California, White Sands, and the British-Australian Guided

Missile Range at Woomera, Australia, as well as the numerous astronomical observatories located in the free world. More important, an eastward launching in an approximate equatorial orbit would impart about one thousand additional miles per hour to the orbital speed, a gain ensured by the eastward rotation of the earth. Cape Caraveral, furthermore, provided an opportunity to launch over a 5,000-mile stretch of the Atlantic Ocean and thus minimize the hazards to human life; if, after the rocket burned out, the booster case did not disintegrate or burn up, it would fall harmlessly into the sea.

Most of the arguments for supporting a satellite project rested throughout on the premise that relatively minor modifications of existing rockets would suffice for the launches. In the ensuing discussion White House advisers weighed the dangers of having development of the vehicle interfere with the ballistic missile program and of spending an excessive amount of money on a comparatively unimportant venture. Consequently when on 26 May the Security Council endorsed a satellite program, the recommendation carried two conditions: the peaceful purposes of the undertaking must be stressed, and it must not interrupt work on intermediate-range and intercontinental ballistic missiles.[11] Although the formal memorandum put no specific ceiling on expenditures, the tacit relegation of the project to a secondary role and the vesting of overall responsibility for the launcher in the Secretary of Defense constituted safeguards against extravagant spending on a scientific will-o'-the-wisp. Moreover, some of the $13 million authorized by Congress for the IGY might well go into the satellite program and thus lighten any monetary burden on the Department of Defense. The endorsement in effect scaled down the project to far smaller dimensions than its staunchest advocates had hoped for, but the Academy proposal had at least escaped outright rejection. With all official obstacles now apparently removed, a start on the great experiment had to wait only for a decision about which launching system to use and who was to take charge.

The choice of launching plan lay with Assistant Secretary of Defense Quarles, for in a secret directive of 8 June Secretary Wilson delegated to him "responsibility for coordinating the implementation of the scientific satellite program within the Department of Defense." Informal conversations between Waterman and Quarles during April and May had produced a tentative agreement on a division of labor whereby the DoD was to provide the rocket, launching facilities, and that somewhat vaguely all-inclusive commodity known as "logistic support," while the IGY National Committee took charge of choosing the experiments and of devising and procuring the satellite instrumentation, the satellite shell, and the scientific equipment for the observation stations. The Science Foundation was to be the official inter-

Homer J. Stewart.

mediary between the Academy and the Pentagon. If, as Kaplan's letter of 6 May to Waterman implied, the USNC would have preferred to keep control of the entire program, the National Research Council nevertheless welcomed the proposed arrangement, since the Academy could not with propriety serve as a governmental operating agency. And Quarles' Advisory Group on Special Capabilities was now in a position to examine the alternative satellite plans and select the most suitable. Quarles himself had named two of the eight-man group, and the Army, the Navy, and the Air Force had each nominated two, but which service had chosen which members was never revealed.[12]

Generally called the Stewart Committee for its chairman, Homer J. Stewart of the Jet Propulsion Laboratory at the California Institute of Technology, the group included Charles C. Lauritsen, an eminent physicist and, like Stewart, a professor at Cal Tech, Joseph Kaplan, chairman of the United States National Committee for the IGY, Richard Porter, consultant for Advanced Developments to the General Electric Company's Missile Division, George H. Clement of the RAND Corporation, Clifford C. Furnas,

Chancellor of the University of Buffalo, J. Barkley Rosser, rocket ballistician and professor of mathematics at Cornell University, and Robert McMath, professor of astronomy and head of the McMath Hulbert Observatory at the University of Michigan. None of the committee saw the paper submitted to the Security Council or Nelson Rockefeller's appraisal of the urgency of having the United States be first to launch a satellite, but every member knew of the Soviet's interplanetary communications commission and was aware of what that might foreshadow. Instructed, however, to bear in mind that noninterference with ballistic missile development was imperative and to regard the satellite program as purely scientific rather than politically significant, the Stewart Committee logically could be expected to put less emphasis upon an early performance than upon the scientific contributions that would derive from the system chosen. Any successful launching between 1 July 1957 and the end of 1958 would meet the IGY objective; what the satellite relayed back might well be more important than whether it began its orbiting in the autumn of 1957 or in 1958. In actuality, the staff of the Academy's National Committee inclined to attribute to the ad hoc group as a whole more interest in applied than in basic science—more concern, in short, for the solution of a technical problem than in the accumulation of fundamental scientific knowledge.

Stewart, Lauritsen, Furnas, and Clement, together with Lieutenant Colonel George F. Brown and Alvin Waggoner of the Guided Missiles Committee staff of the DoD Research and Development Division, spent two days in late June at the Jet Propulsion Laboratory in Pasadena and a third day at the Air Force's Western Development Division. The full committee met at the Pentagon on 6 July. After a morning of briefing by Quarles' staff and a summary of "pertinent satellite studies" given by a RAND representative, the committee met at the Naval Research Laboratory and heard the proposal entitled "A Scientific Satellite Program." The Air Force and Army presentations came the next day. On 8 July, after an executive session, the committee visited the Glenn L. Martin plant to see the work layout on the Viking rocket, and to discuss with Martin engineers its adaptation to a satellite vehicle. On 9 July came a long conference with Army missile experts from Redstone Arsenal during which Wernher von Braun spoke for two hours about Orbiter. From 20 to 23 July the committee hammered out a revised version of its first draft report, and on the 29th three members conferred with Quarles about a third draft.[13] No critic then or later could accuse the group of making snap judgments.

In the interim, while the USNC urged the Science Foundation to secure money from the Bureau of the Budget even before a $10-million appropriation bill for the IGY passed Congress on 30 June, Waterman briefed the President's assistants and further discussed procedures and responsibilities with Quarles. On 27 July at a morning session with the Foundation director and Under Secretary Herbert Hoover, Jr., as spokesman for the Department of State, the President agreed to announce the United States satellite program on the 29th without waiting for the Stewart Committee's choice of a launching system. Intelligence reports suggested that to postpone release of the news would be to risk having the U.S.S.R. make a similar announcement first. Joseph Kaplan immediately dispatched a letter to Sydney Chapman, president of CSAGI, telling him that the United States was about to act on the CSAGI recommendation.[14] As Eisenhower's press secretary James Hagerty wanted to dramatize the occasion, Waterman and—in the absence of President Bronk—the Academy's executive officer Douglas Cornell acquiesced in Hagerty's holding a secret preliminary press briefing at the White House on the 28th, followed by the public announcement the next morning and a press conference with TV and radio coverage that afternoon. The scientists stipulated, however, that no word of the President's decision must leak out before they could notify the Secretary of CSAGI, Marcel Nicolet, in Brussels. To enable ICSU to hear the news at the same time as the American public, Neil Carothers, Waterman's assistant, caught a plane to New York that evening and turned over to a London-bound friend a letter containing the announcement for Dr. Bronk, who would transmit the message to Nicolet on the 29th. Until 28 July, Dr. Waterman later estimated, not more than a hundred people had any inkling of the well-kept secret.

Hagerty pulled all stops in his arrangements. Without mentioning the subject of the advance briefing, his invitation to White House correspondents merely hinted at an important revelation to come. At the opening of the two-hour session on the 28th, he described the satellite project briefly, whereupon the reporters rushed for the exits and the telephones outside, only to find the doors of the conference room locked. Hagerty was making sure that no one sprang the story before the next day's official release of the news. He then called upon Waterman, Cornell, Alan Shapley, and Athelstan Spilhaus of the IGY committee to elaborate and answer questions about the plan. To the amusement of the scientists, that afternoon one irate newspaperman protested that he was a crime reporter; how was he to handle this scientific stuff? [15]

Hagerty's statement of the next day ran:

On behalf of the President, I am now announcing that the President has approved plans by this country for going ahead with the launching of small earth-circling satellites as part of the United States participation in the International Geophysical Year. . . . This program will for the first time in history enable scientists throughout the world to make sustained observations in the regions beyond the earth's atmosphere.

The President expressed personal gratification that the American pro-

Principals at White House press conference, 28 July 1955, announcing U.S. participation in the IGY satellite program: left to right, seated, Alan T. Waterman, James C. Hagerty, S. Douglas Cornell, and Alan H. Shapley. Standing, J. Wallace Joyce and Athelstan F. Spilhaus.

gram will provide scientists of all nations this important and unique opportunity for the advancement of science.

If some of Hagerty's audience were discomfited at the confident tone of phrases like "will for the first time in history" and "will provide . . . unique opportunity," the afternoon TV and radio session enabled Cornell, Waterman, Spilhaus, and Shapley to allude to possible difficulties and still reassure listeners that the program involved no danger to world peace. Releases put out by the Secretary of Defense and jointly by the National Academy and the Science Foundation contained further specifics. The Pentagon statement summarized the fruits of space probes in the past, the advantages of a satellite which would circle the earth once every ninety minutes, and the plans to have the three armed services contribute their technical skills while other scientists determined the nature of the experiments to be undertaken. Bronk's and Waterman's release explained that although the Department of Defense would "provide the required equipment and facilities for launching the satellite," the Academy and the National Science Foundation were sponsoring the program. A single succinct paragraph outlined the reasons for undertaking the project:

The atmosphere of the earth acts as a huge shield against many of the types of radiation and objects that are found in outer space. It protects the earth from things which are known to be or might be harmful to human life, such as excessive ultra-violet radiation, cosmic rays, and those solid particles known as meteorites. At the same time, however, it deprives man of the opportunity to observe many of the things that could contribute to a better understanding of the universe. In order to acquire data that are presently unobtainable, it is most important that scientists be able to place instruments outside the earth's atmosphere in such a way that they can make continuing records of the various properties about which information is desired. In the past vertical rocket flights to extreme altitudes have provided some of the desired information, but such flights are limited to very short periods of time. Only by the use of a satellite can sustained observations in both space and time be achieved. Such observations will also indicate the conditions that would have to be met and the difficulties that would have to be overcome, if the day comes when man goes beyond the earth's atmosphere in his travels.[16]

Four days later the Moscow press announced that the U.S.S.R. would put a satellite into orbit during the IGY. Furthermore, at the meeting of the International Astronautics Federation in Copenhagen, according to a story in the New York *Herald Tribune* of 3 August, a distinguished Russian physicist declared that the Soviet satellite would be launched in 1957 and would be much bigger than any the United States would attempt. Some Americans were alarmed, and some were disdainful about both nations' announcements, but a greater number appeared to be more curious than uneasy. In Dr. Kaplan's opinion, the record was now straight:

The clear recognition of this program as that of the scientists of the nation as gathered into our [The Academy's] Committee and Panel structure, aside from the hard facts of the matter stemming from their conception and intensive work on the program since October 3, 1954, provided, through the CSAGI, the international basis for friendly reception of the program. I was glad to see, in the course of the announcement of the program and subsequent news inquiries, that much good use was made of the material prepared by our Committee, and particularly the material in the program budget document on LPR [long playing rocket]. This gave a good solid basis for our releases and comments.

Fortunately for Dr. Kaplan's peace of mind, he could not foresee that the budget by 1959 would have risen to eleven times the committee's estimate. In the summer of 1955 most Americans wanted above all to know how the United States was going to accomplish this strange undertaking. On 3 August 1955 the Stewart Committee itself had not agreed on the method.[17]

3

SELECTING A SATELLITE PLAN

DAYS of discussion during July left the Stewart Committee divided about whether the Orbiter using the Army's Redstone missile or the NRL satellite scheme based on the Viking rocket would best answer IGY purposes. The Air Force submitted a plan but only as a proposal that might be adopted if neither alternative were acceptable. The Air Force paper, containing nineteen pages of text and twenty-one of drawings and charts, was an elaborate dissertation on the scientific information attainable from a 150-pound satellite if launched by the Atlas rocket then under development. Lieutenant Colonel R. F. Lang of the Air Research and Development Command in making the presentation explained that his service had chosen 150 pounds as a minimum payload; the Atlas would be powerful enough to put hundreds of pounds, "even thousands of pounds," into orbit. The rocket held every promise of success and of growth possibilities, would use proven components and only two stages, would possess a low g factor, and would offer the advantage of simplicity of design. The Air Force, moreover, could supply full logistical support, and a preliminary estimate put the over-all cost at $16,350,000. But, Lang admitted, even if a minimim satellite were made part of the Atlas program, interference with the ICBM development would be inescapable because of competition for facilities, propulsion sources, and skilled personnel. Furthermore, the first launchings of Atlas–B were scheduled for January, February, and March 1958, by which time four launching stands and four assembly buildings would be available. At best that date would leave an uncomfortably narrow time margin for the IGY, and were the Air Force to take on responsibility for a satellite, flight tests of the rocket might have to be further postponed.[1] Homer Stewart remarked in 1963 that such caution had in actuality proved needless, but, at the time, the Air Force had every reason to fear delays in its ICBM developments. Under the circumstances the committee shelved the Atlas–B scheme.[2]

Atlas–B launch.

The Army–ONR Orbiter proposal, dated 1 July 1955, offered a design considerably modified from the original scheme von Braun had tendered as a "Minimum Satellite Vehicle" in September 1954. As an alternative to the clusters of Loki rockets, von Braun and his associates had adopted a suggestion of the Jet Propulsion Laboratory to use Sergeant solid-fuel rockets reduced in size to power the second, third, and fourth stages. Either configuration should be satisfactory. Other changes recommended by JPL and incorporated in the new version consisted of refinements in the engineering of the upper stages. Equally important, tacitly recognizing the "million-in-one chance" of locating a small body in the vast expanses of space by relying solely on optical equipment, the Orbiter team had added a provision for electronic tracking: it might employ either NRL's "light, low-powered transmitter" and large radar directional antennas or else a device, upon which the Army's Diamond Ordnance Fuze Laboratory was working, which would use Lincoln Laboratory radar and dipole antenna modified with a central coil. The Orbiter scheme still called initially for a satellite of only five pounds, although the launcher would be powerful enough to carry a much heavier payload. Estimated costs ran to $17,700,000, of which $6,400,000 would be for eight Redstone missiles for the first stage.[3] However inadequate this and the still smaller Air Force figure would look by 1958, both were less unrealistic than the $9,734,500 Kaplan mentioned to Waterman as sufficient for ten satellites and ten launchers.[4] In the spring and summer of 1955 only guesses were possible about expenditures.

The NRL presentation made no attempt to estimate total costs. While the proposal was explicit about the scientific advantages its orginators envisaged, and in explaining the mathematical formulas upon which they based their calculations, data about the launching vehicle were more general than detailed. The plan called for a three-stage carrier capped by a twenty-inch-long instrumented cone as the satellite, but the text did not describe the mechanism for spinning and firing the second and third stages in flight or the device for separating the satellite from the third stage. After discussing the importance of analyzing the flight path in order to minimize possible errors in projection, the written proposal offered two alternative vehicle configurations. One comprised a M–10 Viking first stage and two solid-propellant stages, the other a liquid-liquid-solid combination consisting of the two-stage M–15 rocket and a solid-propellant third stage.[5]

The M–10 was a modification of the Navy's sounding rocket which had reached to over one hundred miles altitude in seven flights between 1950 and June 1954. Stripped of fins and equipped with the General Electric Company's Hermes power plant, the new version, NRL asserted, could attain an altitude of 216 miles with a tangential velocity of 5,060 feet a second. Much smaller than the Redstone, the M–10 would have a four-foot diameter, a forty-foot length, and a dry weight of 2,250 pounds. The Glenn L. Martin Company had spent two years in studying the design and had prepared detailed drawings of what would be the smallest available vehicle that could serve as a first stage for a satellite. The small size made the M–10 easily transportable by motor vehicle and reduced to a minimum the amount of logistic support needed. The Atlantic Research Corporation had designed the two solid-propellant stages, and both the Martin Company and NRL had scrutinized the plans and considered them sufficiently "conservative" to permit developing and testing within the required time. This combination would put a forty-pound instrumented payload into orbit at a perigee of 216 miles unless an error in projection of the solid-propellant stages occurred; an 0.88° error would reduce perigee to 150 miles. A post-cutoff control system similar to that already in use in the Viking would tilt the carrier over to a predetermined angle to enter the circular orbit.

Adoption of the second configuration would enable the vehicle to carry a forty-pound instrumented satellite into an elliptical orbit with a 303-mile perigee. The M–15 was the M–10 Viking with an Aerobee-Hi liquid-propellant second stage. As early as 1949 the Army's Bumper-Wac, a liquid-fueled second-stage rocket fired by a V–2 in flight, had established the practicability of that type of propulsion system. Under NRL scientific direction, the Aerojet General Corporation had spent two years in developing the Aerobee-Hi; the first flight was scheduled for August 1955. Since the angular precision required to produce an orbit varies inversely with the altitude of projection, one major advantage this combination promised was the higher altitude of projection. At 200 miles, analysis showed, the tolerance would be 0.67°; so that a very precise orienting mechanism would be necessary; at 300 miles the tolerance would be 2.0° and thus permit use of simple and more reliable orienting equipment. With the more efficient flight path made possible by the Aerobee-Hi, the two liquid stages would burn in succession; the second stage would coast to orbital altitude, be oriented, and then spin and fire the third stage. A disadvantage of the M–15 configuration, however, lay in its needing more time than would the M–10 for design and test of the second-stage controls. Whereas the latter would be ready to launch a first satellite two years after work began on the program, the M–15 would have to have an additional six months. "Both configurations," the statement noted, "could be carried forward simultaneously, since the M–10, the most expensive stage, is common to both. The liquid-liquid-solid combination appears to offer more in the way of growth potential."

As for contractors to produce the vehicle, the authors drew attention to the Glenn L. Martin Company's nine years of experience in the design and production of Vikings and the company's "many design and performance studies of satellite systems and components." The Aerojet General Corporation was ready to supply the Aerobee-Hi for the second liquid-propellant stage, and the Atlantic Research Corporation had submitted designs for ARC solid-propellant second- and third-stage rockets. For "system contractor," the agency which would have primary responsibility for the entire program, the recommendation was that the Naval Research Laboratory take charge because of its long-term interest in and familiarity with atmospheric phenomena, radio and radar propagation, optics, radio astronomy, and upper-air research using rockets. The Army Map Service should handle the geodetic measurements and provide the optical tracking instruments, while the National Committee for the IGY should coordinate the geophysical measurements. NRL was ready to supply radio tracking equipment called "Minitrack" which John Mengel and Roger Easton of the Laboratory had devised.

The description of the scientific devices for the satellite and the means of relaying data to recording stations on earth were the most impressive features of the NRL memorandum. Miniaturization, today a commonplace of technology, was a novelty in 1955. The Laboratory's proposal, however, hinged on it. A satellite casing weighing eight pounds would carry miniaturized instruments weighing ten pounds for accumulating scientific data, tiny batteries weighing twelve pounds, Minitrack equipment weighing two pounds and consisting of a miniature electron-beam vacuum tube and a crystal-controlled radio receiver with a hearing-aid-type amplifier to respond to instructions sent by powerful radars at the ground stations, and two pounds of telemetering equipment to transmit information back to earth, bringing the total weight to thirty-four pounds. Although solar cells instead of conventional batteries would save weight, the NRL team discarded the solar-cell source because its dependability was not as yet established. In actuality, solar cells were later incorporated into *Vanguard I*. Investigation of the possibility that the temperature of the satellite might interfere with the proper operation of the instruments indicated that the temperature change would not exceed 10° C as the satellite moved from the day to the night side of the earth and that equilibrium temperature could be kept to 10° C by such simple means as coating half the casing with thick lead-base white paint and leaving the other half of the surface unpolished aluminum.

In an appendix recommending scientific experiments to undertake in a first satellite the NRL team listed five conditions it had imposed upon itself: (1) the information should be significant and obtainable in no other way; (2) the instrumentation required should be of proven design and (3) it should weigh less than forty pounds; (4) the experimental data must be communicable by telemetry; and (5) the experiments must be applicable to a satellite with an equatorial orbit. On the basis of those criteria NRL scientists proposed for the two first experiments to use instruments that could determine the distribution of hydrogen in outer space and could show whether or not a ring current encircles the earth in an equatorial plane beyond the ionosphere.

Two Lyman-alpha detectors in the satellite would serve for the first experiment, one detector to measure the intensity of the ultraviolet radiation emitted by atomic hydrogen from interstellar space, the other less sensitive detector to measure radiation coming directly from the sun. The two together were expected not only to reveal the density of neutral and ionic hydrogen in space within ninety million miles of the earth but also to furnish data on the motions, densities, and sizes of streams of particles ejected from the sun. Physicists and astronomers suspected that the streams affect cosmic ray intensity, the aurora, and magnetic storms. For the second experiment a highly sensitive magnetometer, an instrument for measuring magnetic elements, would serve. Placed in the satellite, it could detect the presence of a ring current and gauge the intensity of the earth's magnetic field, provided the altitude of the satellite were known within 1.7 miles. The Minitrack system could probably meet that proviso. But to interpret the results of the hydrogen density experiment and to correct the roll and tumble rates for the magnetometer, the attitude of the satellite had to be known for any given moment of flight. The method outlined for obtaining that information was to measure by means of a miniature electron-beam vacuum tube the angle between an axis in the satellite and the earth's magnetic field. The device would be actuated from the ground, allowed a minute for warm-up, and would then transmit for three minutes when the satellite was above the ground receiving station. A second signal from a ground-based transmitter or else a timer could turn the instrumentation off and thus lengthen its useful life.[6]

The appendix of the memorandum wound up with the statement that the instrumentation was already available and would need only slight modification to fit into the container, the "pot," as it came to be called. A total of $110,000 should cover the costs, including about $10,000 for instrumentation. Construction of the telemetering ground station would come to about $30,000, and salaries of the men at the receiving station would add about $100 a day. The actuating station would be the most expensive item—apart, that is, from the launcher. The powerful "classified" radar would cost "a few million dollars," unless it were possible to borrow the unit. The Orbiter cost analysis had slid over these elements.

Accompanying the NRL text with its careful exposition and mathematical calculations were graphs and a photograph of "The Double-Axis Phase-Comparison Angle-Tracking Unit," a forerunner of the Minitrack. The Minitrack in fact would be essentially identical with the unit pictured

*Breadboard display of miniaturized components
for Vanguard satellite.*

except for operating frequencies and antenna configurations. A phase comparison angle tracking system in effect determines the angle of arrival of a radio signal by measuring the difference in length between two radio paths from the signal source in the satellite to each of two receiving antennas located on the ground at a known distance apart. The system would provide three coordinates of satellite position and three vectors of satellite velocity, plus accurate time of transit at each ground station once during each pass of the satellite. Since radio transmissions from a subminiature transmitter in the satellite would be feasible at any time of day and under all normal weather conditions excepting severe local thunderstorms, the Minitrack would furnish complete tracking and position information throughout the life of the equipment. A diagram showing the layout of a single Minitrack ground station from the antennas to the recorder clarified the textual explanation.[7]

For scientists eager to get reliable data on the ionosphere and interstellar space beyond, the NRL presentation could hardly fail to have a strong appeal. The Orbiter and the Air Force propositions also discussed the scientific benefits they could offer, but neither was as specific about how its measuring and tracking schemes were to work. Furthermore, the five-pound satellite of Orbiter was manifestly a far less useful research tool than NRL's bigger, more elaborately instrumented payload. The relative merits of the launchers were another matter. The Stewart Committee, several of whose members were interested in basic research as well as in engineering technology, faced a difficult choice, its difficulty doubtless heightened after the presidential announcement and the attendant publicity. The meeting on 3 August, at which the committee prepared its formal recommendations, took place without Professor McMath of the University of Michigan Observatory. Illness kept him from attending, a fateful circumstance if, as rumored, he later declared that he would have voted with the minority.[8] Had he been on hand to do so, the minority might easily have become the majority, for when three men endorsed the NRL proposal and two the Orbiter, the remaining two, explaining that they were not guided missile experts, chose to go along with the numerical majority; if the split had been three to three, the fence-sitters might have landed on the other side. Homer Stewart admitted privately in 1960 that some of the ad hoc Group disliked the idea of using a booster that was a modification of a Nazi Vengeance missile developed by German engineers; an American IGY satellite launcher should be an American product. But, Stewart added, that line of reasoning had had little bearing on the majority's decision.[9]

The report sent to Secretary Quarles on 4 August laid down first the general conditions that would have to be observed to attain success no matter which design was selected. These conditions included a satellite kept below fifty pounds and a perigee of the orbit of not more than one hundred fifty to two hundred miles above the earth. Whatever the launching system adopted, some development work would be necessary and so would present the risk of interference with military programs. Yet if properly carried out, the project would produce long-term military as well as scientific benefits. Only clear, undivided administrative responsibility could fulfill the objective. "Great caution is imperative to insure that existing techniques, existing contractors, group skills and facilities be used." To forestall diversion of resources, top-level control would be essential: "otherwise additional and unnecessary delays will be inevitable." The cost would probably total about $20,000,000, and would be more were "full advantage . . . [not taken] of existing programs, facilities, and reasonable logistical support." The over-all undertaking should embrace two phases, the first realizable before the end of 1958, the second a long-term scheme which could ensure a higher orbit and a payload of as much as a ton. For the second phase an ICBM booster, such as the Air Force Atlas–B, should be used and the Air Force put in charge of the program, but, in view of the uncertainty about whether the ICBM development could keep up to schedule, the Stewart Committee regarded specific recommendations about Phase II a matter beyond its competence. For the IGY program the choice narrowed down to Orbiter or the modified Viking and a payload.

In arriving at recommendations for the first phase, the committee, after agreeing upon the practicability of putting up "anything" during the IGY, considered nine factors: (1) the minimum payload and altitude that could provide "something useful"; (2) duration of the orbit; (3) tracking requirements; (4) the growth potential of the equipment, meaning its chances of leading to more sophisticated scientific and military devices; (5) maximum use of available facilities and skills; (6) minimum delay to military projects; (7) maximum scientific utility; (8) broad national interest; and (9) over-all economy during a five-year period. The group tacitly equated "broad national interest" with success in a launching achieved without interference with ballistic missile programs and at a cost the economy could stand and over which the public—or at least Congress—would not boggle. The relative dollar price of one proposal as over against the other and the effects on the American economy did not occupy the committee long. The estimates submitted were, after all, only estimates and those given in the NRL memorandum covered only the Minitrack system. Still, part of the committee believed the bigger, heavier Orbiter vehicle would cost considerably more than the Viking combination, and excessively high costs for a first satellite would imperil the chances of a continuing program.[10] Nor did duration of orbit net much committee attention. Ironically enough, long orbiting life would prove to be one of *Vanguard I*'s notable features; not only is it still circling the earth and expected to remain in orbit for at least two centuries, but its telemetry system transmitted radio signals for over seven years. In August of 1955 the remaining six factors posed more important and more controversial questions.

The choice of the committee majority was the M–15 with an Aerobee-Hi liquid-propellant second stage, even though that configuration would require six months more to develop than the M–10 with two solid-propellant stages. The proponents of the M–15 argued that, despite its small size and less than 30,000-pound thrust, it offered better performance and more reserve margin than Redstone with its 75,000-pound thrust. Doubts about the efficiency of the latter as a satellite booster in fact took George Clement to Huntsville on 28 July to look for himself. He concluded that the Redstone was too heavy for the purpose and, even if stripped of some of its boilerplate, would probably still be relatively unsatisfactory.[11] Instead of

Orbiter's four stages, the second and third consisting of multiple clusters and deemed proportionately less reliable than single-rocket stages, the three-stage design of the M–15 gave the NRL proposal another advantage. The smaller launcher would also require less logistic support and thus prove more economical for continued satellite use after the IGY was over. The modifications needed for the Viking appeared to be well within engineering capabilities and, because of the probable availability of a new General Electric rocket engine, would avoid any interference with weapon projects. Unless an ICBM rocket motor replaced that of the Redstone in the near future, Orbiter, in the opinion of three committee members, was less likely to succeed than the Viking combination.

Stewart and Furnas took exception to that interpretation. In their view the Redstone booster had more power and flexibility and fewer development problems; as part of an active weapon program it was already under test and had range facilities at its disposal which would minimize interference with military programs. Viking was a greater risk, if only because the margins of error allowed at each stage were so narrow as to be "at the limit of current engineering knowledge"; to correct malfunctions would take precious time. Since the IGY was to end in December 1958, the main question was, in Stewart's words, "what could be done with things that already had some development history and could work at a small level without starting from scratch?" [12] The critics did not emphasize the fact that the Navy would have to beg for space at the Florida missile center, if, as NRL proposed, the Viking were to be projected eastward in order to take advantage of the earth's rotation to heighten the vehicle's velocity. Certainly neither NRL nor the Stewart Committee majority when making its decision envisaged fully the delays and general wear-and-tear that would spring from the Navy's having to negotiate for a launching pad and blockhouse, and then fight for testing time at Cape Canaveral, not to mention the design changes in the rocket which the safety rules at the Air Force base required.

Viking's opponents did, however, point out that the adequacy of the second- and third-stage fuels was not firmly established. Increased performance of the second-stage Aerobee-Hi by use of unsymmetrical dimethylhydrazine, UDMH for short, had not yet been put to the test, although in the Air Force's Bomarc motor the Aerojet Corporation had mixed UDMH with JP–4, the standard kerosene fuel for jets, and had run one test of a twenty-five percent UDMH and seventy-five percent aniline-furfural mixture. Even the time schedule for fuel tests was uncertain. The design of an attitude stabilization system was on hand, but the attitude control system would have to undergo careful testing with the new fuel. The third-stage solid fuel called for ammonium perchlorate dispersed and fused in a polyvinyl chloride matrix which the Atlantic Research Corporation had tested in small motors but not in large. To get an efficient structural design in an end-burning configuration with a long burning time might take longer than the project could afford. For the spin-stabilized third stage, moreover, NRL had not presented any analysis of how to forestall low frequency oscillations; those might prevent attainment of an accurate orbit.

To these objections defenders replied that equal uncertainty applied to Redstone's fuel, and the UDMH system looked like "a very straightforward engineering procedure." Inasmuch as the Glenn L. Martin Company had a feasible approach with commercially available components for an attitude control system, the development presented no greater difficulty than that for a conventional missile autopilot system. Spin stabilization had to be worked out for Orbiter also; the problem was as hard for one as for the other. Although Redstone testing facilities already existed at Patrick Air Force Base, using them for testing the Orbiter satellite vehicle would interfere with ballistic missile programs as much as Viking satellite tests would. The military and scientific projects should be divorced from each other as completely as possible. The Viking was a research rocket, not a weapon. Without saying so in so many words, supporters of the NRL plan apparently felt that under the aegis of a research laboratory manned by civilians, the satellite venture would avoid much of the military flavor likely to permeate it were it directed by the Army Ballistic Missile Agency. NRL had had long experience in atmospheric research, had a deserved reputation for meeting time schedules, and at the moment had no high-priority work afoot.

More significant, the Redstone satellite plan offered relatively little growth potential for future space exploration, whereas NRL's opened up a variety of possibilities, a committee appraisal that was to prove sound. While that consideration doubtless loomed larger to some members than to others, no one dismissed it lightly. On the other hand, the very innovations in the Viking-based design presented greater risks of delay than did the more orthodox features of Orbiter.[13]

Unanimity prevailed on one point: the superiority of the NRL tracking system and indeed all the NRL satellite instrumentation. What the committee would have liked to recommend, Clifford Furnas said later, was the use of the Army's rocket and the Navy's "pot" of instruments. The hitch about that procedure was the intensity of service rivalries. No military service was willing to give "personnel, money—or even information, at times—to a project for which some other branch would get the most credit. We finally decided," wrote Furnas, "that breaking the space barrier would be an easier task than breaking the interservice barrier." That conclusion reached, it was a "toss-up between the Army and the Navy plans." [14]

The committee was at pains to suggest needed improvements to both plans. Orbiter ought to include better satellite instrumentation; eventually the Redstone motor should be replaced with a liquid-oxygen and gasoline motor, and, in the interest of reliability, possibly the second stage should also use a liquid propellant. NRL should schedule a far larger number of tests of components and rocket stages preliminary to attempting a satellite launch and should consider substitution of a Sergeant-type, solid-propellant third-stage rocket for the Arcite-type proposed. Those recommendations, if reflecting committee doubts about the adequacy of both plans as they stood, implied that modifications might make either one satisfactory. Both the Redstone and the NRL teams were ready to adopt the committee's suggestions. Quarles, faced with a five to two decision in favor of the NRL proposal, put the matter up to his Policy Council. The council, after listening to the arguments of its Army and Navy members, voted to postpone a final choice for two weeks.[15]

During that interval, in response to vigorous protests from Major General Leslie Simon of the Army Ordnance Corps, who insisted that misinterpretation of facts had prejudiced the case for Orbiter, Quarles asked the Stewart Committee to reexamine the Navy-Redstone plan. General Simon's memorandum, dated 15 August 1955, asserted:

> The substitution of the 135,000-pound North American rocket engine for the current 75,000-pound engine in the Redstone missile is a less complicated operation than the design of a new Viking missile. There is greater assurance the Redstone with 135,000-pound motor will be available within a 2-year period. Actually the first orbital flight of this improved Redstone motor can take place in August 1957. Using three scaled Sergeant high speed stages, a payload of 162 pounds can be placed in an orbit with a perigee of 216 miles. Payload can be traded for excess velocity. There is sufficient excess velocity to place a 100-pound payload on the moon.[16]

In actuality the launcher thus described was never built.[17] Simon continued:

> The development problems confronting the Viking development make it obvious that the probability of success within the IGY is low. This conclusion is reinforced by looking at the development times of major missile programs already completed. Such programs are rarely completed on the originally predicted dates and require 5 years at the minimum and usually run for approximately 8 years.
>
> The improved Redstone 75,000-pound performance permits improved payloads at orbitable altitudes. The following table was computed with 900 feet per second excess velocity and with existing propellants:

Perigee altitude (miles)	Payload (pounds)
300	6
216	18

The first orbital flight for this configuration can be scheduled for January 1957 if an immediate approval is granted. Since this is the date by which the U.S.S.R. may well be ready to launch, U.S. prestige dictates that every effort should be made to launch the first U.S. satellite at that time. Although this time scale is dependent on Sergeant or Loki clusters, the engineering feasibility has been approved by four competent agencies.

The satellite missile does not interfere with the Redstone missile program because the program is in process of being turned over to the Chrysler Corp. and because the designers and planners are completing their work with the Redstone missile. A new program is therefore necessary for adequate utilization of the talent available. If a new and challenging project is not soon placed at Redstone Arsenal, the loss of key personnel will jeopardize the successful completion of the Redstone missile project. Therefore the scientific satellite program will strengthen rather than weaken the Redstone missile project.

In view of the fact that Army Ordnance can provide heavier orbital payloads with shorter time scales and with greater assurance of success, and that the Naval Research Laboratory is already heavily committed to the Aerobee-Hi development for the IGY, it is obvious that the Naval Research Laboratories will be better employed instrumenting properly the large payloads (100 pounds or more) which can be made available.[18]

The allusion to Aerobee-Hi development referred to improvements in sounding rockets which the IGY committee was anxious to use in continuing probes to supplement data from satellite experiments.

That General Simon's calculation of the perigee and payload in relation to the velocity attainable in the new high-powered version of Orbiter would correspond almost exactly to those achieved in January 1958 by the Army's Explorer satellite demonstrates the soundness of most of his predictions.[19] His gloomy prophesy that rejection of Orbiter would cause "the loss of key personnel" at the Arsenal and thus jeopardize completion of the Redstone ballistic missile, on the contrary, turned out to be erroneous. To some ears it sounded like a piece of special pleading smacking of high-pressure political maneuvering. His dictum about what NRL should be doing did not sit well with Navy men who had faith in the Laboratory's special talents. And NRL and the Martin Company were to prove him mistaken in declaring that the development of any missile required "5 years at the minimum": Vanguard cut that time in half.

When Captain Samuel Tucker, the director of NRL, learned that the Stewart Committee was again to review Orbiter, he discussed the situation with Milton Rosen. Rosen suggested enlisting the aid of Vice Admiral John Sides of the Office of the Chief of Naval Operations in getting a second hearing for NRL also. The sympathetic Admiral Sides advised them to put their case to Admiral Robert P. Briscoe, Deputy Chief of Naval Operations. Briscoe at once volunteered to talk to Paul "Red" Smith, Assistant Secretary of the Navy for Research and Development. Indignant at the Army's attempt to snatch from the Navy its fairly won victory, Smith arranged for a second NRL presentation to the Stewart Committee. By then cost estimates for the Viking-based plan were ready, putting the figure for the launchers at $10.4 million and the over-all cost at $12 million.[20]

For the second NRL hearing Milton Rosen prepared a concise summary of the new features the Laboratory had introduced into its initial proposition. Substitution of a Sergeant-type third-stage rocket in keeping with the committee's recommendation necessitated a reduction in the weight of the satellite from 40 to 21.5 pounds for a 303-mile-perigee orbit, but specifications and performance data furnished by the Thiokol Chemical Corporation showed that a twelve-inch-diameter scale of its T–65 rocket which was then in production would meet NRL requirements. Laboratory calculations, moreover, indicated that the launcher thus modified could achieve a final velocity of 27,730 feet per second with a ten-pound payload at a 200-mile-perigee orbit, while computations undertaken by the Glenn L. Martin Company put the velocity at 28,350 feet per second, twelve percent above the required speed. The test schedule was now to include three firings of the first and second stages separately, a "large number" of the new third stage, and three of second and third stage combined before attempting a satellite launch at all. "The Laboratory is confident," read the final sentence of the summary, "that the first satellite can be launched eighteen months from the start of the program." [21]

Rosen had originally put the time needed at thirty months—an exactly accurate figure as events later showed—but the Martin Company believed a year and a half sufficient. Under pressure to pare down his estimate, especially as "we were fighting for our lives against a competitor who confidently said he could do the job in eighteen months," Rosen succumbed, accepting against his better judgment the more optimistic figure.[22]

To fortify faith in the Laboratory's capacity to meet so tight a time schedule, Rosen appended to his memorandum copies of telegrams and a letter from the four major industrial firms with whom NRL would expect to deal. A wire from the Thiokol Chemical Corporation on 22 August promised delivery of an enlarged solid-fueled T–65 rocket in nine months from receipt of a contract. A similar nine-month delivery guarantee on the power plant for the first-stage booster came from the General Electric Company. The Aerojet General Corporation wired that it was prepared to make a first delivery of Aerobee-Hi engines for the launcher's second stage eleven months after signing a contract, but as heat transfer difficulties were to be expected if UDMH fuel was to be used, "some development firings" would be necessary to test the propellant.

The longest communication was from the Glenn L. Martin Company, producer of the Viking. The company's executive vice president stated: "We see no reason why it should not be possible to put a satellite in being in approximately 18 months provided the program is well defined and effectively managed." After reiterating that "the mission to be performed must be defined clearly," he added: "both government and industry must understand clearly the part each is to play in the program execution." A curiously admonitory tone pervaded the message, but it concluded on an encouraging, if slightly boastful, note: "We recognize that we have systems management experience in addition to the specific experience gained from Viking, but"—and this passage in Rosen's copy of the document was underlined in red pencil—"whether we are called upon to manage the program or to provide the airframe alone, you can be assured that we will support the program in the aggressive fashion necessary to achieve a satellite at the earliest practical date." [23]

These assurances from reputable industrial firms, particularly in regard to delivery dates, made an impression upon the Stewart Committee, since the time element, tellingly stressed by the Army general who spoke for Orbiter, now appeared to be about equal in both propositions. It left the strength of the pro-Viking arguments unimpaired—the sophistication of the instrumentation and the electronic tracking system, the miniaturization of parts, and the adaptability of the design to more elaborate spacecraft in the future. So the earlier verdict in favor of the NRL satellite stood.[24]

Oral word of the decision reached the Laboratory and Redstone Arsenal some seven weeks after the first presentations to the Group on Special Capabilities and more than a fortnight before the Deputy Secretary of Defense officially notified the Secretaries of the Army, Navy, and Air Force on 9 September that the Navy was to be in charge of a joint three-service program.[25] The Army officers promoting Orbiter were incensed. General John Medaris of the Ordnance Corps privately labeled the rejection "a boondoggle." Commander George W. Hoover of ONR was incredulous that knowledgeable men who had listened to von Braun and examined the sheaves of detailed drawings prepared by him and his staff could have accepted an alternative consisting of "a blueprint of a pencil-shaped vehicle" many parts of which existed as yet only in the imagination of its authors. Frederick C. Durant III, one of the original backers of Orbiter, recounted that in the post-Sputnik era—after the failure of the first attempted Vanguard satellite launching and the subsequent success of the Army's vehicle—a National Academy official remarked ruefully that "one of the major reasons for the Army's losing out was von Braun's lousy presentation." Durant dubbed that explanation "odd" in view of von Braun's gifts of lucid exposition and persuasiveness.[26] Conceivably in the course of his two-hour speech to the committee in July, the German rocket expert appeared so to exaggerate Orbiter's technical capabilities as to raise doubts in the minds of his audience.[27]

If only because several members of the IGY National Committee had served in the Navy during World War II and had acquired high respect for the caliber of Navy research, one might speculate as to whether they had greater confidence in the scientific environment at the Naval Research

Laboratory than in that at Redstone Arsenal. Nothing, however, suggests that National Academy preferences, if known, influenced the Stewart Committee's choice. Outside the Department of Defense a number of scientists assumed that a big factor in the decision had been the relative security classification of the two boosters: Redstone, intended to be not only a test vehicle but a weapon in the American defense arsenal, would have to carry a secret tag, whereas the modified Viking would not; and secrecy would run counter to the IGY plan of sharing information with other nations. That idea was a misconception. The committee had known from the first that the guidance and control systems of any satellite vehicle would have to be military secrets; security considerations consequently had played little part in committee discussions.[28]

At NRL the staff was at once elated and frankly surprised. To those who knew most about the competing proposals the chances had looked minimal that the slim forty-foot Viking booster with a third stage of only partly determined design could win against the powerful sixty-nine-foot Redstone into which nearly four years of development work had already gone. Although the magnitude of the task to be accomplished by the Laboratory in three years might well have induced a touch of stage fright, excitement over the challenge submerged every doubt. Then and later, the NRL team attributed its victory to the quality of the scientific data the plan promised to produce and to the prospects it held out for future advances in rocket technology and space exploration.[29]

4

GETTING THE LAUNCH PROGRAM STARTED

SOME confusion and a number of conferences occurred at the Naval Research Laboratory during the fortnight before and the three weeks after Deputy Secretary of Defense Reuben Robertson issued his September 1955 memorandum outlining the military departments' respective obligations under the joint satellite program.[1] With technical responsibility assigned to the Navy, it was a foregone conclusion at the Pentagon that the Laboratory, under the administrative aegis of the Office of Naval Research, would direct the project. As it happened, the Secretary of the Navy waited until 27 September formally to designate ONR as administrator and not until 6 October did the Chief of Naval Research officially notify the Laboratory that it was to take charge.[2] ONR, a Navy officer once remarked, served the Laboratory chiefly as a post office—transmitting funds, inquiries, and, occasionally, directives from higher authority. But an organization chart drawn up somewhat later shows how many echelons of authority stood above NRL. Although a long and distinquished record in research and engineering had won a measure of independence for the Laboratory, its chiefs were still bound by orders from ONR and higher ranking officialdom. Small wonder that uncertainty reigned during late August and September 1955.

As time was precious, the military and civilian directors of the Laboratory felt obliged to make a number of tentative decisions well before they received notice of their authority to proceed. They had to decide how large a proportion of their staff to assign to the satellite program, who was to head it, how much freedom of action he and his principal assistants should have, in consultation with them how to refine the budget estimates and initiate procedures to ensure a prompt and orderly flow of money, how to handle publicity in keeping with whatever security classification the Navy imposed upon the project, and, most urgent of all, what ground rules to lay down in negotiating a contract with an industrial company for design and production of the launch vehicle.

The Laboratory's key men realized that the primary contract would have to deviate in some respects from the routine kind whereby the company selected through competitive bidding undertook to meet requirements for a prototype by building, test-firing, and making successive changes in a series of models. In this case, time forbade that orderly procedure. NRL, acting for the government on an intricate task involving a number of federal agencies, would have to exercise constant supervision over the producer and be free to rewrite specifications if necessary as the work progressed. Performance of mutually interdependent systems, designed in part by a process of extrapolation and analogy, rather than solely by proven, predetermined specifications, would spell success or failure in this novel experiment. Hence, before making any other decisions, the day after NRL learned of its victory, Captain Samuel Tucker, the Laboratory's director, invited two Martin Company executives to his office to hear the good news and talk over plans. For from the inception of the NRL satellite plan its originators had taken for granted that, were their proposal accepted, they would again be working with the engineers who had built the Viking. At Rosen's suggestion, during early summer the Martin Company had prepared several analyses of aerodynamic problems affecting the project and had made a formal presentation to the Stewart Committee. The committee indeed had indicated that its selection of the NRL scheme was predicated partly on the assumption that Martin experience would expedite design and production of the launch vehicle.

Two additional Vikings were already on order for continuation of NRL's work on upper-atmosphere and guided missile research, including a study for the Air Force on electromagnetic wave propagation in rocket flames; cancellation of that commitment and diversion of Vikings 13 and 14 to use as satellite launching test vehicles should get the program off to a good start. The Martin Company consequently was sure of its favored status.

When the two Martin vice presidents appeared in Captain Tucker's office on August 25, their confidence in their impregnable position became evident. Tucker, Wayne Hall, NRL's deputy director of research, and Rosen all felt the chill in the atmosphere when the Martin men disclaimed interest in a contract unless the company had full control. As the job was unlikely to be a big money-maker, prestige was the sole inducement. Whereas the public always spoke of the Navy Viking rocket, this time the label must be the Martin satellite project. Equable but firm responses from the NRL men, however, led the manufacturers to agree to a more formal conference on August 31 and to transformation of the existing contract for

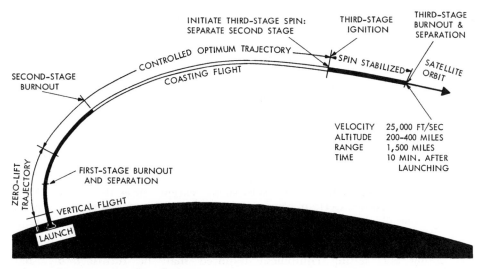

Schematic of Vanguard trajectory.

Vikings 13 and 14 into one releasing the vehicles for the satellite program.[3]

To expedite matters at the meeting held on the 31st, William G. Purdy, chief spokesman there for the Martin Company, came armed with a draft "change-order" of the Viking contract, immediate acceptance of which, completed by filling in a few blanks, would constitute, he opined, a suitable contract for the satellite project and permit work at the Baltimore plant to begin at once. After stating that the government had designated ONR as the "procuring agency" for a program aimed at creating small scientific earth satellites, "the first of which is intended to be operational within approximately eighteen months," the draft change-order declared that the Martin Company as the prime contractor would "carry responsibility for the development, design, manufacture, and test of the complete system." ONR should:

> assign to this program a Project Officer, who shall be the Government representative authorized to commit the Government on contractual and technical decisions. . . . The Contractor shall rely on the Project Officer for the proper coordination of Government furnished services, facilities and materials and for approvals, where required, of Contractor activities. Subsequent to acceptance by the Government of the system specification, changes in the specifications will require joint approval of the Project Officer and the Contractor.

The contractor should then proceed immediately "to furnish the necessary services, materials and facilities" in accordance with the following requirements:

> a) The mission of the system shall be to place a satellite of____pounds and____cubic inches in an orbital plane within____degrees of the

59

plane of the equator and having a minimum altitude in any point in the orbit of____miles.

b) The Government shall furnish the complete satellite package, including instrumentation, antennae and all ground equipment associated with the transmission and reception of scientific information. The configuration of the satellite shall be established by mutual agreement between the Contractor and the Project Officer within 60 days of the date of this letter.

c) The vehicle shall be launched from Patrick Air Force Base, with the Government furnishing propellants and such handling and test facilities, together with the necessary personnel to operate the same, as are not included in the Contractor's responsibilities under (e) below.

d) The Contractor shall develop, design, manufacture and test the complete vehicle, including all subsystems, except those defined by (b) and (c) above. Certain components may be sub-contracted at the discretion of the Contractor.

e) Handling and testing equipment facilities presently available to the Government shall be utilized whenever possible. . . . Additional handling and testing equipment required shall be furnished by the Contractor.

The two Viking rockets on order were to be used to advance the development of the satellite vehicle. Within ninety days of signing this contract the Martin Company would submit a firm proposal which would include a system specification, a test plan, and an estimate of the time and cost required for completion of the entire program. Here, from the manufacturer's standpoint, was a straightforward plan of procedure.

The Laboratory and ONR representatives objected: the scheme relegated NRL scientists to the sidelines and demanded of them a delegation of responsibility which, in light of the Defense Department's mandate to the Navy, was out of the question. A single project officer, corresponding to the company's operations manager, could not make final decisions for the government; any significant departure from the proposal approved by the Stewart Committee would have to have both committee and NRL concurrence, lest the end product fall short of the scientific performance required. Captain Tucker pointed out that the contractor would be dealing with NRL, not with ONR, despite the latter's acting as the Laboratory's agent in drawing up the contract. Everyone present acknowledged the importance of definition of the mission, prompt formulation of operational and logistic plans, definition of mutual responsibility, and centralized management; but the Navy representatives could not establish then and there a clear-cut division of responsibility between NRL and the contractor. Neither could they map out explicit operational and logistic plans as yet, since the Army and Air Force were also involved. For the present, Martin should put into writing its recommendations about "system responsibility," vehicle specifications accompanied by delivery schedules, test firing schedules, firing responsibilities and procedures, launching sites, range requirements, the company's requirements for vehicle instrumentation, and a list of items the government should furnish. NRL itself would provide telemetering and tracking equipment and take charge of its operation. While the company was drafting its recommendations, the Laboratory staff would prepare specifications, undertake to integrate them with Martin's, and define the Scientific Officer's duties. A second meeting should take place on 8 September, by which time both parties ought to be able to arrive at a mutually satisfactory modus operandi. When the conference broke up, the only matters on which both sides had reached agreement were the necessity of further meetings, the "Confidential" classification for most components of the vehicle, and NRL's providing the satellite package.[4]

During the week following this exchange, Laboratory specialists by twos and threes carried on a series of exploratory negotiations. Because a reliable engine for the third stage of the satellite launcher was a pressing need and the Thiokol Chemical Company, under contract with Army Ordnance, was working on a solid-fuel T–65 engine at Redstone Arsenal, on 1 September James M. Bridger and Alton E. Jones of NRL and two Martin engineers met with a Thiokol Company official and a man from the Arsenal to discuss arrangements whereby Thiokol could use the Arsenal facilities in designing and developing a third-stage power plant for the NRL project. The group concluded that before proceeding further NRL and Martin must agree on specifications and the Thiokol Company must map out a research and development program on a schedule acceptable to the Arsenal. A few days later, while Milton Rosen, Bridger, and General Electric Company engineers talked over preliminary plans for a new GE engine for the first-stage rocket, NRL telemetry experts, headed by Daniel G. Mazur, flew to Patrick Air Force Base in Florida to examine the instrumentation available there, to inquire about safety rules and test schedules, and to explain what facilities the Navy hoped to have the Air Force supply. Mazur brought back some discouraging news: owing to high priority military projects scheduled for tests during the next eighteen months, space for satellite vehicle tests would be hard to arrange; safety rules to forestall danger to ground installations and to Atlantic shipping lanes would require a power-cutoff device in each stage of the rocket; and, as the base lacked the type of telemetry equipment wanted for the satellite and its launcher, NRL would have to either supply and man its own or else modify its original plans.[5]

Meanwhile in Washington the Navy, accepting the name suggested by Mrs. Milton Rosen, christened the satellite venture Project Vanguard. At the same time consensus at NRL and ONR ran that the Laboratory should establish a separate Vanguard unit which would include the entire Rocket Development Branch and some men from other branches and divisions. Captain Tucker and Edward O. Hulburt, the Laboratory's research director, then selected John P. Hagen to head the project. Although he had had no part in drafting the satellite proposal, he was the obvious choice, Hulburt averred. Superintendent of the Laboratory's Astronomy and Astrophysics Division, Hagen possessed unquestioned stature as a scientist. If his slight build, his bespectacled brown eyes, his mild-mannered professorial mien and utter lack of facade led outsiders at first into thinking him a misfit for a difficult administrative job, his scientific knowledge and his imperturbability in dealing with temperamental prima donnas quickly impressed high-powered industrialists as well as his own staff. In spite of later complaints that he was incapable of "dynamic leadership," he commanded the respect and affection of his colleagues. Indeed one of his team attested that a more forceful chief would probably have cracked under the strains the job imposed; Hagen's patience and tolerance were essential to seeing it through.

Under Hagen the intense, hard-driving rocket expert Milton Rosen was to be technical director. The youthful-looking Homer E. Newell, named "Science Programs Coordinator," would work closely with the National Academy's IGY panels. Hagen early asked to have Thomas Jenkins, the Laboratory's Deputy Comptroller, as his budget officer; Jenkins, a soft-spoken young man, thenceforward wore two hats, one for the Laboratory as a whole, one for Vanguard. Other appointments came from the Rocket Development Branch, the progenitor of the Vanguard plan: Leopold Winkler as Rosen's engineering consultant on problems of mechanical design, James M. Bridger on rocket propulsion and control, John T. Mengel on satellite tracking, and Daniel G. Mazur on telemetry and data reduction.[6] The sad-eyed, big-eared Winkler, with his air of resignation to human vagaries, was as even-tempered as the short, slightly rotund, humorous Mazur was explosive in the face of crisis; tall, lanky, handsome Bridger tended to the laconic but at times was as emphatic as Mazur; Mengel, heavily built, sandy-haired and square-jowled, displayed the quiet self-possession of a man sure of his knowledge, yet ready to learn from others. All four shared a passionate interest in the satellite program and each brought a special expertise to it.

How to map out financial procedures was more difficult than setting up an informal organization for Vanguard. On 8 September Captain Tucker issued a new budget estimate raising to $28.8 million the $20-million figure submitted to the Stewart Committee in mid-August. Deputy Secretary of Defense Robertson's directive of 9 September reiterated the National Security Council's decree that the satellite venture must not interfere with military programs of higher priority, instructed each service to "provide for the immediate implementation" of action on its share of the joint undertaking, and specified that all information intended for public release must first be

John P. Hagen, Director of Project Vanguard.

submitted to the Office of Security Review for clearance. But the paragraph covering budgetary arrangements merely noted that "the Navy Department will manage the technical program with policy guidance from the Assistant Secretary of Defense (R&D) and will provide the funds required . . . with the understanding that reimbursement will be made as soon as funds can be made available from other sources." The decision as to how much money, when, and from what sources was left in abeyance.

Cryptic, unsigned notes on an informal conference held shortly thereafter among representatives from Quarles' office, ONR, and the Laboratory revealed the extent of uncertainties. Asked about what the Laboratory would expect to receive from the National Science Foundation, Admiral Furth said, "NSF is out," whereupon Charles Weaver of Quarles' staff observed that the Foundation could supply some funds. When the Admiral suggested that the Navy simply do what it could for $20 million, Weaver reminded him that that sum, the figure which the National Security Council had accepted, was not binding and in any case did not include the cost of satellite instrumentation. Apparently everyone assumed the availability, free, of a good deal of expensive equipment already in existence and a number of services—test and launching facilities, for example, powerful radar, and telemetry instruments, as well as Army Engineer Corps construction, maintenance, and operating teams for the tracking station network.

A decade later perusal of the early discussions of costs might awaken in the reader the suspicion that a lack of candor prevailed among participants fearful lest realistic figures cancel the entire program. But the Vanguard comptroller, the person most familiar with the financial problems, saw "guesstimates"—all too often too low—as inescapable: there were no cost data to draw upon applicable to a satellite program. In 1955 no one had ever built a multistage launcher except as a vertical sounding rocket; no one had tried out an electronic tracking system for an artificial earth-circling body; and no one could predict accurately the expenses of developing reliable instrumentation capable of functioning for days in space. The vagueness that shrouded early financial planning was a handicap to the men at NRL responsible for getting work started. For a first allotment of money they had to wait until December 1955 and in the interim had to draw on the Laboratory's small Naval Industrial Fund.[7] In mid-September they were still engrossed in contract negotiations with the Glenn L. Martin Company.

For both NRL and the Martin Company resumption of the contract discussions proved an uncomfortable experience. So far from achieving a meeting of the minds, the exchanges on the 8th of September and again on the 12th resulted in a sharpening of differences. As foreshadowed in the earlier session, the principal items of contention were two: who was to have overall "systems responsibility," and whether unilateral changes in specifications were to be permissible. NRL insisted that the Laboratory itself must hold the reins, the Martin men that the company must, once a firm contract had spelled out requirements. William Purdy declared that Martin's authority must embrace not only fabrication and testing of the launching vehicle but also control of supporting facilities and operations and of ground instrumentation for tracking and telemetering. In addition, Martin demanded freedom to "make design decisions within a system specification," an equal voice with NRL in defining the system specification, and, once that was fixed, no changes without company acquiescence. NRL replied that it would not accept "a closed-door policy on government-directed changes." The impasse on the major issues was so complete that NRL representatives wondered briefly whether they dared risk looking for some other prime contractor. Obviously that step was no solution. The Martin Company had the Viking experience and had prepared satellite launching studies during the spring and summer. A switch to another company would require the express approval of Assistant Secretary Quarles and the Stewart Committee and would certainly be costly in time. One young NRL engineer later said bitterly of the Martin negotiators, "They had us over a barrel, and they knew it." [8]

The irritation was mutual. The Martin representatives were apparently following the line laid down in the company's front office, but they also felt that concurrence with NRL's stated policy would subject them to needless harassment—a succession of Laboratory scientists invading the Baltimore plant and breathing down the necks of company executives and engineers, or worse, a series of peremptory directives from Washington making impossible demands and causing endless delays in what would be at best a very tight time schedule. Purdy and Robert Schlechter, who had both worked harmoniously with Milton Rosen and other members of NRL's Rocket Sonde Branch in developing the Viking rockets, were particularly irked at the reluctance on the part of the Laboratory to delegate authority to a contractor who had proved trustworthy and efficient in the past. But because the contract was important to the company as well as to the government agency, the men confronting each other across the conference table agreed to put to one side the question of specification changes and division of responsibility until Rosen and Schlechter had reviewed and revised the outlines of specifications and test schedules so as to permit drafting and signing of a "letter of intent." The letter of intent should serve as a preliminary to a final contract that would pin down reciprocal obligations and financial arrangements.

By 15 September the air had cleared slightly. The associate director of the Laboratory informed the Martin representatives that an "initiating contract" was ready for their scrutiny, that John Hagen was to be the administrator of the project with Rosen as technical director, that a redefinition of the mission had relaxed the eighteen-month requirement for launching but called for a "program of quality" at reasonable cost; the company was to submit cost figures the next day. Discussion then produced a specification outline that both groups considered a sound basis for "initiating system definition." Either group could request further meetings on specifications, and the same rules were to apply to subcontracted items as to Martin's. Disposition of the earlier contract for Vikings 13 and 14 was to be a separate negotiation. The flight test program presented by the Martin engineers, however, met with criticisms from James Bridger and Daniel Mazur because it was a minimum one with inadequate backup in some areas, and because more exact information on trajectories was needed, since Martin's computations assumed at each stage mass ratios which were far from optimum. Mazur pointed out, furthermore, that tests without the right telemetering instrumentation would be useless and that the Air Force Missile Test Center's safety doctrine was likely to impose severe "instrumentation weight problems"; design of the controls in the second stage in particular would require close attention to keep them within weight limits. John Mengel added that instrumentation for the third stage of vehicles spending substantial time above 200 miles of altitude was another problem to consider carefully.[9] Those objections notwithstanding, the way now looked clear to signing a workable letter of intent within a few days.

The agreement as it took form labeled the Glenn L. Martin Company "the supplier of the launching vehicle" which was to orbit a satellite that "will enhance the prestige of the United States." The satellite was to be trackable by radio, optical instruments, or both. The contractor was to prepare the vehicle for launching, to suggest ways of helping the scientific phases of the undertaking, and by mid-January 1956 to have a complete time schedule worked out. The NRL Scientific Officer, the director of the Laboratory, or whomever he designated, namely Milton Rosen, was to have access to all data and, if he wished, to attend all conferences. Within forty-five days of signing the letter of intent the company was to submit to the Scientific Officer its recommended specifications and within sixty days have them in finished form. Any changes the government deemed essential were to go to the contractor in writing and any proposed by the latter must have Rosen's written approval. The company must notify him of the effects of changes on deliveries and costs. The detailed design, if up to specifications, need not have express government endorsement. The payload was to weigh 21.5 pounds. The minimal altitude of projection at the final stage must be 300 miles to ensure a perigee of more than 200 and an apogee of less than 800 miles. Eastward launchings from Cape Canaveral would give an inclination of the orbit to the equator of about 30°, but all vehicles were to be capable of achieving a 45° angle. Until a regular reporting system was evolved, the contractor was to submit semimonthly progress letters to NRL. A final contract containing the definitive specifications for the vehicle and a careful enumeration of government and contractor obligations should replace the letter of intent within 120 days.[10]

By 23 September, when the Navy and Glenn L. Martin officials signed the initial contract for Vanguard, Martin had set $13 million as its probable costs, although, in keeping with standard governmental procedures of "incremental financing," the sum written into the contract was only $2,035,033.[11] Despite the unresolved questions of specifications, both parties thought the agreement reasonable.

Gratification at NRL evaporated a few days later: the Martin Company had just won the Air Force contract to build the airframe of the mighty Titan missile and was in the process of assigning a majority of its best men to start that job in Denver. Vanguard would have to get along without Purdy's hand at the helm in Baltimore and without other experienced engineers who had helped develop the series of Vikings step by step.

While realizing that the Titan job would be bigger and more remunerative for the contractor, people at NRL were dismayed. So also were members of the Stewart Committee who had counted on Martin's putting, not some but all, its top-flight designers and engineers to work on Vanguard. The IGY staff at the National Academy too felt uneasy. Indeed all the principal sponsors of Vanguard believed that Martin was relegating the satellite project to the role of poor relation; denied a high priority within the Department of Defense, it had now become a second-string project in the manufacturer's books also. Martin executives denied the allegation when they learned of it; they were fully aware, they declared, that the satellite project was important and had assigned to it a fair proportion of their engineering talent. They had chosen Elliott Felt, Jr., to head the Vanguard team at the Baltimore plant. Though new to the responsibilities of top-level management, he was an able, hard-headed engineer who had worked on controls for Vikings 9 and 10. Still, as Titan would require 1,000 designers and engineers and Vanguard some 300, the company was obviously going to have to spread its manpower resources thin. Whether or not the Titan contract did in fact seriously impede progress on the satellite vehicle perhaps mattered less than NRL's conviction that it was being shortchanged, for that belief shook the Laboratory's confidence in the contractor's good faith.

Equally troublesome from NRL's standpoint was Martin's new "get-tough" policy, based apparently on the company's ambition to enlarge its position in the industry and taking the form of a determination to yield as little as possible to extravagant demands of impractical scientists and government bureaucrats. Proud of having wider experience than any other firm in the country in designing and building rockets and ballistic missiles, the company considered itself entitled to speak with authority. At the first informal meeting in Captain Tucker's office on 25 August, Martin executives had outlined their position, but the NRL group had believed they would modify it when they understood more fully the purposes of the program. Unhappily, the contretemps between company officials and government scientists in Washington assumed new proportions with every passing week of the autumn.[12]

In essence the clash arose from conflicting philosophies, although not at first recognized as such. The "definition of mission" as stated in the initial contract fixed the Martin Company's commitment: to produce a workable launcher for an earth-circling satellite before the end of 1958. A hard enough job in itself, in Elliott Felt's view it should not be complicated by extraneous inquiries from the company's customer into why something worked or did not; if it failed, investigation should be confined to finding something that would function properly. The scientist, Felt observed wryly, always wanted to know the whys of success or failure; the practical engineer was intent on getting performance. It was for that reason that Martin had hoped to have a commissioned Navy officer, instead of a scientist, put in charge for the government. Felt was obviously oversimplifying, using the term "scientist" loosely, but the difference in attitude of mind he was pointing to was unmistakable before the end of 1955. NRL, as a scientific research body, contended that the study of failures could and should be as constructive as chance success; learning the whys was part of the job, an essential process in building up knowledge to draw upon for future work. Where the manufacturer was ready to try shortcuts and rely upon empirical data, the Laboratory held out for a more carefully analyzed approach. Hence NRL's insistence that it must not only supervise plans but monitor their execution. The ensuing controversies, intensified at times by the greater freedom the Air Force allowed the contractor on Titan, would wax and wane over the next two years but never entirely subside.[13]

In beginning their common task, both NRL and Martin recognized the importance of determining promptly the dimensions and estimated weight of the vehicle, the aerodynamic loads, and the thrust and efficiency of the engines. The first- and second-stage rockets had to have more powerful propulsion systems than either the Viking or the existing Aerobee-Hi carried. For the solid-fuel third stage even some of the preliminary requirements were still uncertain. And the weight and configuration of each stage would affect those allowable in the other two as well as overall performance. Had perfect harmony obtained between the Vanguard team in Washington and the men at the Martin plant, the problems confronting them in their race against time must still have been gargantuan. Until the major specifications were fixed for each part of the whole launcher, work at the drawing boards could not progress far.

A first necessity was to decide upon the optimum distribution of weight within the vehicle. Taking as a point of departure the calculation of Joseph Siry of NRL that a launch vehicle of about twenty thousand pounds gross weight would produce the required orbital velocity, at the end of September Bridger and Richard L. Snodgrass met with Robert Schlechter and other Martin engineers to discuss how much weight to allot to each of the three stages. The give-and-take at that session illustrates the difficulties of handling technical questions involving a number of unknowns. Although a large part of the vehicle was of standard design, recourse to extrapolation would be necessary at many points. When one of the Martin representatives argued that since nobody as yet had ascertained the minimum weights possible for various parts, a proposed "optimization parameter study" would be a waste of time, Snodgrass suggested that a rough guide based on preliminary estimated weights would save trouble, for if scrutiny of the three-stage system indicated that the second stage must not weigh over 2,000 pounds, the designer would not consider a control system suited to a second stage weighing 4,000 pounds; a second study based on detailed design weights could follow and a final study based upon hardware weights could modify that. While the GLM men doubted the feasibility of keeping to Siry's 20,000-pound figure, they agreed, since they had as yet no surer calculation to work from, to start with it in the endeavor to achieve a vehicle that could reach a peak altitude of 299.94 miles with a tangential velocity of 26,420 feet per second, an excess over orbital velocity of 2,890 feet per second. Martin engineers later pointed to their skillful employment of the analog technique as one of the significant contributions Vanguard designers made to the nascent science of space exploration.[14]

Concurrent with the weight optimization studies, work on the desired flight trajectory had to proceed, even though the computations had to be tentative until the weights of the vehicle stages were fixed and although a new plotting would have to be prepared for every flight. By means of special "calculation programs" employing both digital computers and Reeve Electronic Analog Computers (REAC), Navy scientists and Martin aerodynamic engineers arrived at attainable flight paths with an allowance of a six-degree deviation which would satisfy range safety requirements and still permit the transmission of scientific data to ground stations throughout the life of the satellite's batteries. In plotting the optimum path from the initial vertical lift at takeoff to the horizontal altitude of the orbiting satellite, analytical work undertaken during the early winter at the Naval Ordnance Research Center at the proving ground in Dahlgren, Virginia, provided valuable data to start with. "The Vanguard 3–D [three-dimensional] trajectory program," wrote a Martin engineer in 1960, "was probably the most important single tool of the project in defining and solving the design and flight problems." In the later design specifications covering the trajectory, the only change from those of the preliminary version was, first, the extension of the apogee to 1,400 miles and eventually the removal of any limit. In actuality, the first Vanguard satellite to attain orbit would have a 2,460-mile apogee but still proved able to transmit signals receivable on earth.[15]

Anxious to get subcontractors started on the job, in mid-September 1955 Martin with NRL concurrence had entered into negotiations with the General Electric Company to build the power plant for the first-stage booster. Reaction Motors Inc., the maker of the Viking engine, was working on a motor with a 75,000-pound thrust, but Rosen and Bridger thought that much power excessive for Vanguard and, furthermore, they dared not wait to see how the only partly developed RMI engine would perform. What GE had to offer looked like a better choice.[16] Martin itself planned to supply the tankage and vector control actuators, but requested GE to furnish a self-contained unit which was to include the thrust structure, gimbal ring, engine components, and engine starting equipment. To use tooling methods originally employed in building Vikings, the diameter of the cylindrical tanks constituting the rocket casing was to be forty-five inches; the configuration of the GE engine must mate with that dimension, just as engine inlet pressures and temperatures must integrate with the Martin tankage and pressurization systems. The new GE X–405 engine was expected to achieve 27,000 pounds thrust with a specific impulse of 254 seconds at sea level. The term *specific impulse* is a measure of a rocket engine's efficiency—the higher the specific impulse, the better the engine. It is equal numerically to the pounds of thrust an engine produces by burning propellant at a rate of one pound a second. Delivery of the first of ten X–405 engines was to be on 1 October 1956. General Electric engineers accepted the terms, but Rosen asked for more detailed specifications on propellant utilization and insisted that GE supply a complement of spare parts and an itemized price list. With those changes incorporated, the purchase order from Martin to GE was signed on 1 October.[17]

A contract for the second-stage engine was harder to arrange. The preliminary specifications based on modifications of the small Aerobee-Hi rocket called for hypergolic propellants—fuels that ignite spontaneously when mixed—and regenerative cooling of the thrust chamber. Those features together with pressure feeding of the fuel into the thrust chamber promised to provide reliable means of starting the engine at an altitude of about thirty-six miles, even though igniting a pump-fed engine at a high altitude was a still untried system.[18] The Aerojet General Corporation, maker of the Aerobee-Hi, had expressed interest in a contract in mid-August, but the Bell Aircraft Corporation also had claims to consider because of its extensive experience in developing liquid-fuel engines, notably for the Nike antimissile missile. While Martin approached both contractors, Bridger took counsel with the Air Force and Redstone Arsenal. The Air Force put the competitors on a par; the Army gave Bell's performance a slightly higher rating. When the two companies presented their proposals for second-stage propulsion systems complete with structure and tankage, Martin engineers judged Bell's 8,000-pound-thrust, pressure-fed, integral-tank design technically superior to Aerojet's 5,000- to 7,500-pound-thrust, turbopump system. But Aerojet's request for a chance to prepare a pressurized integral-tank design postponed a decision.[19]

While Martin was drawing up a purchase order to the Minneapolis-Honeywell Company for the development of the rocket's guidance and control system, Aerojet won the engine contract, partly, at least, by putting in a bid of $1.03 million, less than half Bell's more carefully figured price

Vanguard engines: rear, GE's first-stage engine;
foreground, Aerojet General's second-stage engine.

of $2.675 million. As things turned out, the costs of the second-stage engine package exclusive of the thrust vector control actuator and its hydraulic tanks would exceed $4 million. The development of the entire second stage, with the complex of equipment it had to accommodate, came to be one of the principal *bêtes noires* of the project during the next two years. Yet eventually it would constitute one of Vanguard's notable contributions to the design of reliable spacecraft.[20]

In the autumn of 1955, however, it was the third-stage rocket that appeared to be the most formidable part of the undertaking. In August the Thiokol Chemical Company had seemed the most likely subcontractor and, with Redstone Arsenal collaboration, had prepared an exploratory study of how to meet the Martin-NRL tentative specifications. But at the end of September a Thiokol spokesman had declared that Martin's demands went beyond the present state of rocketry art. Thiokol could meet the weight specifications for metal parts, but the required impulse would need twenty more pounds of propellant. He saw no way of producing the slow burning time in solid fuel of the diameter GLM called for, and, he contended, under those circumstances it was impossible to reach the thrust level specified; the safety requirement, moreover, was excessive, higher than that set for aircraft rockets. In reporting upon this gloomy analysis, James Bridger noted half-humorously: *"All Is Not Lost,* however." By increasing the diameter of the case, allowing "more optimum expansion," a less exaggerated safety requirement, and use of the new TRX–217 propellant which, tests indicated, could give a burning rate of 0.092 inch per second instead of the 0.28 inch per second of the earlier design, the solid-fuel rocket wanted for Vanguard should be realizable. Slow burning was important to minimize the jolt caused by a rapid buildup of thrust which might damage the delicate instruments in the satellite. As soon as a compromise was reached on those technical requirements, he concluded, progress should be rapid, provided funds were forthcoming promptly.

Bridger's optimism proved ill-founded. By early November, although Martin increased the allowable weight to 365 pounds, Thiokol declared it could not build a rocket of the type needed unless hydrostatic proof pressure were reduced, the case diameter were increased from 11 to 14½ inches, and 25 pounds were added to the gross weight, or else the allowable burning time of the fuel were shortened. Confronted with Thiokol's ultimatum, Martin and NRL felt compelled to restudy the original specifications and in the interim to seek proposals from four other companies for the third-stage engine. There the matter stood till mid-December.[21]

Organization of the Vanguard teams meanwhile had been taking form at the Martin plant and at the Laboratory. In Baltimore Raymond B. Miller, Jr., was made contracts manager; energetic, thirty-year-old Donald

Markarian became project engineer, with Robert Schlechter, Leonard Arnowitz, Joseph E. Burghardt, Russell Walters, and Sears Williams as his assistants. By the end of October, under this supervisory staff, about fifty designers and aerodynamic, electromechanical, and propulsion engineers were working on Vanguard plans. After the readying of production lines and test facilities began, the number rose to some three hundred. At NRL matters moved more slowly, largely because John Hagen wanted to work out a clearcut division of responsibility between one unit and another and to be sure that the head of each, irrespective of his individual attainments, could work effectively not only with his fellows at the Laboratory but also with his counterparts in other government agencies and industry.[22]

As Hagen himself had to devote much of his time to arranging for collaborative services from other branches of the government, to obtaining money from the DoD Emergency Fund or the National Science Foundation, and to the preparation of memos, speeches, and formal reports, additional administrative staff was necessary. So J. Paul Walsh was transferred from another division to become Vanguard deputy director and planning manager; a young man, Walsh brought to his new assignment the vigor and technical leadership which had marked him early in his career in testing nuclear devices in the South Pacific. Charles De Vore, public relations officer for the Laboratory, took charge of that task for Vanguard during its first two years. At the same time, Rosen engaged a few new men, but most of his then thirty-five man staff was already part of NRL and familiar with particular aspects of its work; he estimated that within six months he would need fifty-three men and within a year sixty-six. In late November he set up a technical board consisting of himself, his three branch chiefs (Bridger, Mengel, and Mazur), his engineering consultant Winkler, Walsh, and James Fleming of the Laboratory's Applications Research Division. Homer Newell, in charge of satellite instrumentation, had no separate staff as such; instead he drew on the special talents of some thirty men working in various divisions of the Laboratory but not assigned directly to Vanguard; later nearly a hundred experts worked with him in seeking solutions to such problems as excessive heating of the instruments.

To compute the orbits of close-in satellites Hagen appointed a committee composed of Joseph Siry of NRL, Gerald M. Clemence and R. L. Duncome of the Naval Observatory, and Paul Herget of the University of Cincinnati. For military liaison officers assigned full time to the Laboratory, the Navy first named Captain B. F. Herold and later Commander Winfred E. Berg; the Army appointed Major John T. O'Hea, the Air Force, Lieutenant Colonel Asa Gibbs. All told, what John Hagen called the "in-house management staff" consisted of fifteen men.[23] In view of the official DoD announcement that the satellite program was to be a three-

Some of Project Vanguard key staff: left to right, Commander W. J. Peterson, planning coordinator; Leopold Winkler, engineering consultant; Homer E. Newell, Jr., science program coordinator; Milton W. Rosen, technical director; Mrs. Lillian M. Campbell, secretary to the director; Hagen; James M. Bridger, Vehicle Branch; John T. Mengel, Tracking and Guidance Branch; and Joseph W. Siry, Theory and Analysis Branch.

service undertaking, the question may arise as to why the Army and the Air Force had only one representative each on the Vanguard staff, and why Hagen did not invite Wernher von Braun, for example, to serve as a consultant. Inasmuch as ABMA in 1956 was making overtures to Hagen to enlist his support of an Army plea to use the Redstone-Orbiter launchers as a backup for Vanguard, there is some reason to suppose that von Braun would have accepted the invitation to sit in on Vanguard councils. But, as Hagen later explained, Army collaboration had been wanting when he needed it and he had concluded that the DoD had never "envisioned a greater degree of interservice participation than we [NRL] were able to generate." [24] As it was, the Army Corps of Engineers and the Signal Corps gave the project unstinting cooperation from the beginning. Officers of both Corps went out of their way to cut through entangling departmental red tape and to open doors that might otherwise have long remained closed to civilians and officers at NRL. And after Vanguard field crews arrived at the Cape, the Air Force in turn tendered all possible help.

The entire Vanguard team at its peak numbered 180 persons, including clerical help and shop hands. During 1955 some jockeying for positions of authority went on and a few complaints sounded in other divisions of the Laboratory that Vanguard appointees were disrupting the rest of NRL, but although friction among members of the team recurred from time to time, the Vanguard group early came to be a tight-knit organization bound together by an exceptional esprit de corps. The excitement and passionate interest in the job extended also to the men not formally assigned to the project but who, whether working full time or only part of every month on constructing and testing the instruments for the satellites, considered themselves and were considered part of the Vanguard team. "Team" was the key word. Set up in a fashion that defies exact charting, the organization functioned as a unit. No one within it or associated with it acted on any matter of importance without consulting with Hagen and his management staff, for every part of the operation affected every other. If that procedure sometimes caused delays, it prevented confusion and trouble at a later time, and it heightened the sense of mutual responsibility that all members of the team shared.[25]

Enthusiasm was put to a severe test during the late autumn of 1955. In the midst of skirmishes with Martin officials, NRL and ONR were struggling to secure from the Air Research and Development Command assurances of accommodations for Vanguard tests at the Air Force Missile Test Center at Cape Canaveral. After Mazur's exploratory trip to the base in early September when the Air Force spokesman there outlined the objections to such a plan, and after Vanguard administrators had seen the full text of AFMTC regulations, Vanguard heads had looked for an alternative site. White Sands Proving Ground, where the Navy had equipment for vertical launchings of sounding rockets, had to be ruled out for the satellite program because of the danger to populated areas from falling burned-out first- and second-stage rockets. Use of the Navy installation at Roosevelt Roads in the Caribbean would not involve that peril but would be prohibitively expensive, and inconvenient as well. Hence a formal request from the Navy for space and equipment at the Florida base went to the Air Force on 2 November.

A month later an official Air Force endorsement approved "in principle the support of Project Vanguard at AFMTC" but warned that new construction would probably be necessary or else joint use of existing facilities. The Air Force could not pay for new installations and, if joint use of existing facilities were attempted, "it is possible that other missile programs will be delayed."[26] Unfortunately investigation revealed "that the Redstone facility is the only facility at AFMTC that is suitable with proper modifications for Vanguard operation." And, as Captain Tucker informed the Chief of Naval Research and Assistant Secretary Quarles on 15 December—

It is now stated by the Army representative at AFMTC that these facilities will not be available. . . . Since the first [Vanguard] launchings are scheduled for October 1956, it is apparent that even if money were available and an adequate priority obtained, scheduled commitments for the initial phases of the program cannot be met by totally new construction. Therefore, in order to meet the Vanguard schedule, funds for construction, suitable priorities, and sharing of existing launching facilities are mandatory.

. . . The Test Center Commander also advises that he must have authorization and funds by 1 January 1956 for any construction necessary to the project.[27]

Although the Laboratory won permission to draw for this purpose on the DoD Emergency Fund and had worked out with the Martin Company a statement of Vanguard test requirements, the road ahead looked thorny: the badly wanted high priority for the satellite project was not forthcoming. Lack of priority meant that the Vanguard team could hope for little consideration in obtaining a launch pad equipped with the necessary pipelines and wiring or the privilege of having Vanguard share a blockhouse with a guided missile project. To build those facilities from scratch would take more time and money than the IGY program could afford. And the question remained: how, without adding excessive weight to the rocket and consequently sacrificing velocity, to incorporate in the launch vehicle the safety devices upon which the AFMTC insisted? The Center's headquarters declared that the first and second stages of the vehicle must have both a power cutoff and a destruct receiver capable of functioning from the first moment of flight, and demanded for each Vanguard trajectory such data as fuel weights at successive intervals from launching through the burnout of the final stage, the predicted number of fragments created by a destruct explosion, their velocity, and the time delay between activation of the firing circuits and the first motion. Altogether, the permission to launch Vanguard from Cape Canaveral appeared to introduce nearly as many problems as it solved.[28]

Negotiations for other facilities, on the other hand, proceeded fairly smoothly. Martin had asked in September and the Navy had arranged to have the Malta Test Facility north of Schenectady reactivated for tests of the General Electric engine. Hagen's staff had endorsed the recommendation of Rosen and Robert Schlechter that static testing of the fully assembled first stage not be attempted at Malta because shipment of tankage and instrumentation from the Martin plant to GE would be expensive and would delay production in Baltimore. Doubts later arose about the wisdom of the decision to dispense with full-scale first-stage static tests at Malta. Schlechter for one believed that they would have saved endless troubles at Cape Canaveral and in the long run have proved an economy, but even severe critics of the quality of the rockets sent from Martin to the Cape point out that equal or worse difficulties might have followed from choosing the other course.[29] Although static firing tests could not be conducted in the heavily populated Baltimore area, NRL agreed to finance the building of other special test facilities on Martin property at Strawberry Point on Chesapeake Bay and to meet the cost of constructing at the main plant an elaborate steel tower to permit assembly and inspection of the entire first-stage rocket in a vertical position. The immovable gantry on a forty- by twenty- by three-foot concrete pad would be disassembled after the Vanguard program was over.

In seeking approval of having the government foot the bills for these "costly" installations, Admiral Rawson Bennett, Chief of Naval Research, felt impelled to explain to the Secretary of the Navy that construction and testing for Vanguard involved problems for which there was no existing precedent. No contractor had ever undertaken this kind of job and no one had the testing facilities needed.[30] Nor did that list include the new devices NRL had to have in its shop. For environmental testing, for example, the Laboratory would need a pressure chamber big enough to hold a thirty-inch spherical satellite in which the pressures could be raised from that of 0.01 millimeter of mercury to that of 1 atmosphere; the chamber should provide means of cooling or heating the satellite by radiation; for vibration tests an electromagnetic shaker capable of giving a 625-pound force, a longitudinal and centrifugal accelerator and a shock simulator for use in testing the spin-stabilized third-stage rocket. Equipment already on hand could be adapted to some of this work. Neither the Navy brass nor the Office of the Secretary of Defense protested over expenditures for testing facilities, even though the cost appeared likely to run to nearly $80,000 for Martin alone.[31]

Dealings with the Army Corps of Engineers also went well when NRL approached the Corps about constructing satellite tracking stations. Before discussing plans with the Army, Hagen, Walsh, Mengel, and Roger Easton, Mengel's collaborator in devising the Minitrack system, concluded that eight stations spaced about 850 miles apart in a north-south "fence" along the 75th meridian from Maryland to Santiago, Chile, would provide a sixty percent probability of making a measurement on each orbit of the satellite at an average altitude of 200 miles. NRL proposed to place a station for testing the system and training field crews at Blossom Point, south of Annapolis, Maryland. The Navy itself would man a radar station at Antigua in the Bahamas to observe satellite performance at the moment of third-stage separation, but requested the Army Engineers to build, maintain, and operate the tracking stations in Cuba, Panama, and South America. The Engineers were willing to cooperate. Selection of the exact sites in foreign territory would have to await State Department approval and the results of an exploratory trip by a team composed of members of the Inter-American

Geodetic Survey, the Army Map Service, and NRL representatives. The survey of possible sites was set for late February 1956.[32]

As 1955 drew to its close, men responsible for the satellite venture could see a good deal of progress. At the National Academy plans for choosing scientific experiments to attempt in the instrumented bird were moving forward under the aegis of an IGY satellite panel.[33] At NRL a well-organized team had been formed to ensure that a satisfactory launcher was available at the earliest possible moment, that a flyable bird was ready, and that tracking and data reduction were as accurate as scientific knowledge and technical knowhow permitted. The comptroller had set up orderly procedures which promised to ease, if not obliterate, difficulties in getting money released for approved purposes. NRL had enlisted the help of Army engineers in establishing tracking stations. The Martin Company had aligned three major subcontractors and had several studies in process calculated to ensure sound design of the vehicle. The chief monkey wrench in the works was still the failure of the Laboratory and the primary contractor for the launching vehicle to reach agreement on final specifications.

5

BATTLE OVER VEHICLE SPECIFICATIONS

THE DESIGN specification for the Vanguard launching vehicle completed on 29 February 1956—thirty-one pages of text and three appendixes—discloses little of the effort that went into drafting the document. As sharp differences of opinion arose, virtually every sentence underwent minute scrutiny by both parties to the agreement, and several amendments elaborating policy had to be added later. The task of fixing the principal features of the vehicle was exceptionally difficult, inasmuch as success would have to depend in considerable measure upon innovative advances in the art of rocket design. But it was "policy considerations" rather than the design of the vehicle itself that underlay most of the conflict between top management at the Martin Company and its counterpart at the Naval Research Laboratory. While arguments between government representatives and industrial executives were—and are—routine in the course of negotiating an important contract, in this case they were peculiarly hard to settle because of the novel character of the undertaking: a combination of research and development with production of an operational vehicle to be built at minimal cost in time and money. And the question of prestige—who was to get most of the credit for success—was ever present, if rarely admitted. The final contract between NRL and the Martin Company consequently bore the date 30 April 1956.

On top of the troubles revealed by the Thiokol Company's study of third-stage specifications, during the winter of 1955–6 the problems of weight, reliability, and engine power in the first and second stages harried Vanguard designers. While both NRL and the prime contractor examined the qualifications of companies competing for subcontracts, both teams explored the possibilities of using new materials, of simplifying test procedures, and of making minor modifications in design to obviate the necessity of major changes. In early December, for example, Kurt Stehling vigorously investigated alternatives. A Bell Aircraft rocket engineer whom Rosen engaged to head the Vanguard propulsion section, he pursued with GE men the feasibility of using ceramic liners for nozzles, plastics such as fiberglass for propellant tanks, teflon for tank liners, and aluminum components for the X–405 engine, provided structural instability could be overcome. These materials not only would be light but would not corrode. Since Hagen and Rosen thought a mixture of seventy percent liquid oxygen and thirty percent fluorine (as a high performance oxidizer) in the first-stage motor might heighten combustion efficiency and increase engine thrust to about 30,000 pounds, they requested the Stewart Committee to sanction an independent study of the proposal, but the committee, which had to approve any significant change in the original plan of the vehicle, turned the request down. Teflon and metal-to-metal glandless seals in valves and lines might have to be used with fluorine oxidizers, and the Martin Company was leery of "fluorine hazards." For the second-stage oxidizer white fuming nitric acid might give better performance than red because of the higher boiling point of white and hence its better cooling properties, but might it not have counterbalancing disadvantages? Should an inhibiting agent be added to reduce tank corrosion, or should the tanks be made of a noncorroding metal at the risk of excessive flexibility in the structure?

Studies of alternatives were time-consuming. And how much time did the Vanguard teams dare invest in a search for the best in design and materials rather than settling for what was at hand? Rosen and Bridger at the Laboratory and Markarian and Sears Williams at the Martin plant, the men upon whom fell the major responsibility for vehicle design, were necessarily wary of the experimental. Days were slipping into weeks and weeks into months and vehicle specifications were still tentative.[1]

As the size and configuration of the 21½-pound payload and separation mechanism would necessarily affect the design of the launcher, Martin had begun in September to lay out plans to accommodate a cone-shaped satellite twenty inches in diameter at the base, a configuration that corresponded to the general description included in the original NRL proposal to the Stewart Committee. But, although a cone would involve lesser weight and heat penalties and be much easier to attach to the nose of the third-stage rocket, by mid-October Laboratory scientists had reluctantly yielded to the wishes of men at the National Academy who argued that a spherical shape would better serve scientific purposes. It was a painful decision to make, since Hagen and his staff realized that use of a sphere would necessitate an increase in the size of the second stage, lengthen the time needed for its design and fabrication, and raise costs.

When the Laboratory notified the company on 1 November that the government-furnished satellite was to be a thirty-inch sphere, Leo Winkler

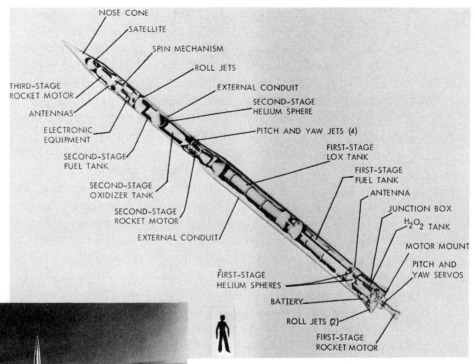

NOSE CONE
SATELLITE
SPIN MECHANISM
ROLL JETS
THIRD-STAGE
ROCKET MOTOR
EXTERNAL CONDUIT
SECOND-STAGE
HELIUM SPHERE
ANTENNAS
ELECTRONIC
EQUIPMENT
PITCH AND YAW JETS (4)
FIRST-STAGE
LOX TANK
SECOND-STAGE
FUEL TANK
FIRST-STAGE
FUEL TANK
SECOND-STAGE
OXIDIZER TANK
ANTENNA
JUNCTION BOX
SECOND-STAGE
ROCKET MOTOR
H_2O_2 TANK
EXTERNAL CONDUIT
MOTOR MOUNT
PITCH AND
YAW SERVOS
FIRST-STAGE
HELIUM SPHERES
BATTERY
ROLL JETS (2)
FIRST-STAGE
ROCKET MOTOR

Cutaway view of the
Vanguard launch vehicle.

Artist's conception
of Vanguard, released
20 March 1956.

81

explained the reasons to protesting Martin engineers: the effects of air drag on a sphere in flight would be easier to measure than those on a cone-shaped body and, as a cone was more likely to tumble and be lost to sight, the Academy's optical tracking program depended on having a spherical body, preferably as large as thirty inches and in no case smaller than twenty inches in diameter. Unless Martin studies proved that a thirty-inch ball was technically impractical, that requirement must stand. After describing NRL's ideas on design of the satellite mounting and separation device, Winkler added that the Laboratory planned to make the sphere of 0.020 aluminum with a 0.001 coating of aluminum oxide. The antennas were to be four wires retracted during ascent of the launcher and released to spring outward after separation of the bird from the burned-out third-stage rocket. Excessive heating of the satellite during the last third of the booster's flight and the first half of second-stage flight could probably be prevented by providing a heat-resisting, disposable, conical shield. Yet even were the cone jettisoned soon after first-stage burnout, it would add weight during the critical minutes after launching. With these strictures in mind, Martin engineers set themselves to computing the effects of twenty-inch and thirty-inch spherical satellites on vehicle performance and analyzing the desired characteristics of a disposable protective nosecone. Winkler remarked that GLM was reluctant "to incorporate many of NRL's ideas" but agreed to consider them.[2]

A sharper conflict occurred meanwhile over the type of telemetry to be used at ground stations, in the test vehicles, and in the satellites during flights. Daniel Mazur had discovered in early September that AFMTC had no pulse-width-modulation/frequency-modulation receiving equipment. If Vanguard were to use minimum size and weight PWM/FM airborne units, new ground antennas and receivers would have to be installed at that base. That would be an expensive undertaking. And, as soon became evident, Air Force officers in charge at the Cape disliked the idea of having special new radar and telemetry brought in for the sole use of any project, and particularly for what a few of them considered a relatively unimportant nonmilitary program. The Martin Company, in turn, believed an elaborate new telemetry system for Vanguard a needless refinement. AFMTC furthermore announced that it would not assume responsibility for the reduction of telemetered data.[3] The Vanguard organization would have to handle and pay for that and set up its own communications center as well.

By November the Vanguard staff at NRL had perceived that, if it was to use the Florida site, it was going to have to convert part of the Cape from a ballistic missile test range to a space vehicle launching range. Even if the Air Force raised no objections, it would be a big job. Vanguard would need new telemetry equipment, a high-precision tracking radar, and "Dovap" antenna. Dovap—an acronym for Doppler, velocity, and position—was a continuous-wave trajectory measuring system using the Doppler effect caused by a target moving relative to a ground transmitter and receiving station. Hope of borrowing Dovap antenna free of charge evaporated when the Army Ballistic Research Laboratory refused to lend its array. Of the FPS–16 tracking radars, the kind NRL experts wanted for Vanguard, only three were in existence, one at the plant of the maker, the Radio Corporation of America; one in Navy hands at Point Mugu, California; and a third in the Army's possession. A new XN–1 model would not be available before July 1956 and a more complex XN–2 not until December and then at a cost of $800,000.[4]

In the vehicle, NRL had planned from the beginning to use for the first stage a pulse-position-modulation/amplitude-modulated telemetry system (PPM/AM) because of its high accuracy; for the second stage, where weight was critical, the lightweight pulse-width-modulation/frequency-modulated system (PWM/FM) would do. The Martin Company wanted to install FM/FM throughout, for it was the standard system with which Martin was familiar and the one in use by the Air Force at Cape Canaveral. Implicitly denying NRL's generally acknowledged preeminence in the field of telemetry, company officials labeled Daniel Mazur's concepts "obsolete" and challenged his contention that FM/FM was heavier; he had not taken into account the weight of the antennas. Mazur, never prone to mince words, declared that Martin's proposals "ignored pressurization problems, flame attenuation problems, and flame pluming problems," difficulties which might seriously interfere with performance of the engine control systems. "GLM," Mazur continued,

> insists on treating a combination of three undesigned and untested rockets as a fully engineered production guided missile which is far from the case. Their assumption that all equipment as prepared in the plant will fly without field preparation is not only fallacious but is not conducive to obtaining data which indicates [sic] how to get the best performance out of the Vanguard vehicles rather than merely verifying that something works.

In his judgment, the company should neither have system responsibility for a complete radio frequency link, nor control procurement of equipment, nor take charge of checkout and the installation of flight units. Martin contended that these were properly its responsibilities; certainly it must have a voice in decisions about what was to go into the vehicles. Indignant at Mazur's disparaging comments, company engineers pointed out that they had had far more experience in managing rocket and missile tests in the field then had anyone at NRL.[5]

As the argument progressed, questions of policy entered into it more largely than did technical considerations, for the crux of the matter, as Rosen saw it, was Martin's determination to "run the whole satellite show." If the Laboratory furnished the telemetering equipment and took charge of its operation, Martin responsibilities would be greatly reduced. A letter written at the end of November by the company's Vanguard contract manager, R. B. Miller, Jr., lent color to that interpretation. He informed NRL that as the "original concept of an over-all Martin responsibility for the design, construction, testing, ground support and launching procedures . . . dictates that the specifications be limited in coverage to general requirements," he could not as yet submit an all-inclusive recommendation on "launching systems specifications." Parts of Martin's specifications "stipulate design objectives rather than absolute requirements" because of designers' doubts about the validity of some performance limitations. Further progress depended on decisions in four areas: configuration of the government-furnished satellite, the third-stage propulsion system, AFMTC facilities requirements, and the type of telemetry to be used.[6]

When Hagen on 12 December submitted to Assistant Secretary Quarles' Policy Council a first Vanguard progress report, the description of the main features of the telemetry system contained no concessions to Martin's ideas. The test vehicles were each to have a PPM/AM AN DKT–7 transmitter in the first stage and a small PWM/FM transmitter in the second along with a radar beacon adapted to use with the best radar available at AFMTC at the time of tests. In the satellite itself telemetry instrumentation would be kept to the minimum necessary to indicate satellite performance and transmit tracking signals to Minitrack ground stations. The Minitrack system called for a radio transmitter within the satellite operating at a frequency of 108 megacycles and having a power output of 50 to 80 milliwatts. The radio signals from this transmitter would illuminate the antennas of the ground station which were so designed as to measure the differences in the path lengths from the source to the individual antennas in the array and thereby obtain the angular position of the radio source in the satellite. To save weight in the satellite, the tracking experts planned to combine use of the Minitrack transmitter and antenna system with the satellite telemetering system. A telemetering command transmitter on the ground would signal to a receiver in the satellite to turn on a telemetering modulator unit and a power amplifier just before the satellite entered the Minitrack antenna beams. A miniaturized transmitter fed with about 0.5 watt power from batteries would then telemeter data to the ground stations over the Minitrack link. One Vanguard contract had already gone to the Elsin Electronics Corporation of Brooklyn for ground station telemetry and another for development of telemetry antennas to the Physical Science Laboratory of the New Mexico College of Agriculture and Mechanical Arts.[7]

As the interim contract with Martin stated that the government would take charge of tracking, as well as communications control and data computing, it seemed logical to the Laboratory to exclude the contractor for the vehicle from negotiations relating to those tasks. Martin's job would be finished when one or more of the six satellite launching vehicles which the company was to furnish had placed a satellite in orbit. But Miller's memorandum indicated that the company felt hamstrung by lack of information about NRL plans and commitments. His comment on the dubious validity of some limitations imposed on Martin designers, moreover, struck an ominous note. When the Stewart Committee met at the Martin plant in late November, members expressed their dissatisfaction with the situation but had no constructive advice to offer on how to improve it.[8]

At a heated session in early December, a company vice president asserted that Martin would never have accepted the Vanguard job had the company known it was not to be the systems manager with full responsibility for the entire project—the vehicle, all supporting facilities, and tracking of the satellite in its orbit. He excepted only the package of scientific instruments and, by implication, the computing center which would translate the data relayed from the satellite to the ground stations. "He views the government," Milton Rosen observed, "as a subcontractor to the Martin Company to provide whatever services the Martin Company needs to do the whole job." In spite of a pointed reminder that in the telegram sent to Admiral Sides in mid-August the company had offered to act either as "systems agency" or as vehicle contractor, the company vice president now labeled the lesser role wholly unacceptable. When asked if Martin had full system responsibility for the Air Force Titan program, he had to say no, only a promise for the future.

In an exchange that followed with Elliott Felt, Rosen not only rebuked the company for failing to submit the weight optimization study but charged the management with obstructing direct communication between NRL and subcontractors. Rosen wanted the design specifications in Martin's purchase order to GE, Aerojet, and other subcontractors made part of the NRL–Martin contract. Felt contended that such a procedure would cause needless complications. He reiterated that, given "design freedom," Martin would quickly produce a satisfactory vehicle. The exchange subtly revealed the different philosophies of the two men toward the undertaking that bound them into a reluctant partnership. The industrialist believed that production of a vehicle capable of putting a satellite into orbit was all that was necessary and all that time would allow, a point of view fortified by the Stewart Committee's recommendation to depend as far as possible on proven design and "off-the-shelf" components. Rosen, impelled by longer range interests, countered: "Performance is not only a design target but a minimum requirement." Both men, however, realized that work on definitive specifications must move faster.[9] By mid-December they were facing the uncomfortable fact that the size and weight of the vehicle might have to be increased if it was to accomplish its purpose.

Everyone associated with the design problem had long recognized the desirability of using a more powerful first-stage rocket, but the GE X–405 engine was the best available for a nonmilitary program and perforce would have to do. Under those circumstances reappraisal showed few changes in the rocket's configuration necessary. Although a larger diameter would have advantages, the finless booster's forty-four-foot length and the forty-five-inch diameter of the cylindrical casing were big enough. Those dimensions would allow for a transition section at the rocket's front end containing a "well" in which the nozzle and part of the thrust chamber of the second stage would sit.

Nor was a weight increase necessary. On the contrary, the specifications signed on 29 February fixed dry weight at approximately 1,789 pounds and propellant weight at 15,499, which, combined with an engine furnishing a specific impulse of 254 seconds, would ensure a vertical velocity of 3,903 feet per second and a horizontal velocity of 4,023.[10] Four factors accounted for keeping the total weight to 17,331 pounds, nearly 1,000 pounds less than NRL had figured earlier: first, the use of aluminum for the tank casings and a thin sheet of magnesium for the cylindrical monocoque spacer between the kerosene and the lox tanks; second, skillful design, notably in the placing of the tankage; third, the miniaturization of parts for the electrical system; and fourth, the reduction of telemetry equipment to a minimum.

Hydrogen peroxide (H_2O_2), decomposed in a catalyst chamber to a gaseous mixture of oxygen and superheated steam, was to provide the energy for turbine-driven pumps to feed the lox and kerosene into the engine. The design put the H_2O_2 tank directly above and off-center from the engine motor but close to the peroxide decomposition chamber. The steam from the decomposition chamber would pass through the turbine and be vented through exhaust nozzles on opposite sides of the airframe. Pivoting of these nozzles would counteract roll motion of the vehicle. Helium gas was to pressurize the fuel tank. Venting the exhaust from the two helium spheres also into the turbine exhaust system would add sufficient thrust to maintain roll control during the period of separation of the first stage from the second. Helium was hard to handle, but its light weight was an asset. Pipes, valves, batteries, and electrical connections were all so located as to be readily accessible through structural doors in the frame. In the process of converting these specifications into reliably performing hardware, a succession of difficulties would crop up and a number of changes would be necessary, such as additional vent valves and a new layout of the plumbing, but few knowledgeable engineers ever found fault with the basic design.[11]

Interestingly enough, although the specifications of February 1956 stated that the government would provide a destruct receiver for the first stage in keeping with AFMTC range safety rules, the table showing the maximum weights allowed for each subsystem did not include an entry for a fail-safe device which, complete with batteries and wiring, would add twenty to thirty pounds to the first stage. The omission was deliberate, based on hopes that the Test Center would relax that requirement for the satellite vehicle. The faith was justified, for in April Milton Rosen sought out Major General Donald N. Yates, commander at Cape Canaveral, and explained why Vanguard designers felt sure that a second-stage destruct receiver would suffice. The general was convinced. Unwilling to imperil the project, he decreed that for Vanguard alone he would authorize the exception.[12]

Troubles over design of the second stage, however, worsened during the winter. And the second stage was critical, if only because it had to carry equipment that controlled the performance of the other two stages. It had to accommodate the guidance system for both the first and the second stages, including a programmer and a three-axis gyro reference system capable of functioning with sufficient accuracy to ensure injection of the third stage into a trajectory leading to an orbit. In addition, the second stage had to carry the mechanism to jettison the protective nosecone which would shield the satellite from the intense aerodynamic heating that would build up during the vehicle's journey through the atmosphere. It had to carry a radar beacon and a command receiver to initiate fuel cutoff and vehicle destruct as required by AFMTC range safety rules. It also had to carry the mechanism for spinning the third-stage rocket and a separation device to detach the third stage from the second after burnout. Although described as "intrinsically a simple unit of pressurized tanks with a rocket combustion chamber at one end," the second stage contained most of the brain directing the functioning of the vehicle.[13]

Since it had early become clear that uprating of the small Aerobee-Hi engine could produce at best only four fifths of the 7,500 pounds of thrust required for Vanguard, a bigger power plant with larger tanks to take more propellant and a redesign of the thrust chamber were essential. Yet "mass ratio," that is, the ratio of the total weight including that of the propellant to the dry weight of the structure and its instrumentation, must be as high as possible; for every pound added to the dry weight would cause a loss of velocity of eight feet or more a second. To provide greater fuel capacity, Martin considered increasing the diameter of the oxidizer and fuel tanks from thirty-two to forty-five inches but discarded the idea in favor of lengthening the stage to more than sixteen feet. Although the elongated tanks could hold some 2,464 pounds of nitric acid and about 897 pounds of UDMH as fuel, even then the resulting thrust would barely suffice to inject the third stage into an orbital path at 300 miles altitude.[14]

Aerojet engineers repeatedly urged adoption of a turbopump system for forcing fuel into the thrust chamber; the scheme had the advantage of saving dry weight by using lighter gauge metals for the tanks than would be feasible in a pressure-fed system. But Martin believed the latter, utilizing heated helium gas as the pressurizing agent, more reliable. And reliability of operation, particularly in starting the engine at thirty-four miles of altitude, was vitally important. Still, the difficulty of devising an economical and workable method of heating the helium and of developing aluminum or magnesium tankage strong enough to withstand the pressures to which it would be subjected apparently inspired Aerojet to announce that the company, if held to Martin's current specifications, could not make a first delivery before mid-April 1957 at the earliest. The flight test schedule, as it stood at the time, called for launching a complete three-stage vehicle, TV–3, on 29 March 1957. Manifestly Martin was facing much the same kind of conflict with its subcontractor as NRL had with GLM. In mid-January Kurt Stehling, appalled by "the nebulous and confusing state" of second- and third-stage design, attempted to discuss the problem with Martin engineers, but they regarded his offer as a reflection upon their competence. They nicknamed him "the hit-and-run engineer." Complaints from company officials about NRL "interference" sounded loud enough to lead Commander Berg, Hagen's special emissary at the contractor's plant and by now the chief dispenser of oil on turbulent waters, solemnly to suggest, tongue in cheek, that Martin withdraw in favor of the General Electric Company as primary contractor: Martin could then become a subcontractor under GE's aegis. A half hour later two Martin executives called John Hagen to assure him that they preferred the existing arrangement.[15] Berg's light touch rapidly reduced tensions.

Time, moreover, was forcing the pace of contract negotiations. When the interim contract between the government and the Martin Company was about to expire in late January, the Office of Naval Research in arranging an extension refused to remit more than a part of the additional $3.5 million requested by the company until the final specifications were completed.[16] Both Vanguard teams, furthermore, knew that decisions on some features of the design would have to evolve as work progressed. The upshot was Martin's acceptance of the NRL version of specifications with a provision for amendments when necessary. As Rosen had promised earlier, the government allowed the contractor design freedom on a number of matters, but carefully spelled out performance requirements and major characteristics. Gross weight of the 16.16-foot second stage was to be approximately

4,770 pounds, of which dry weight was to be 937, propellant weight 3,240, and payload 484, thus achieving a mass ratio of 0.679. Thrust was to be 7,500 pounds, fuel burning time 120 seconds, and specific impulse 278 seconds—which would provide a vertical velocity increment of 2,022 feet per second and a horizontal velocity of 8,339. The oxidizer was to be white fuming nitric acid instead of red as originally called for.[17]

Without describing the type of mechanism to be used, the specifications stipulated that the separation system must jettison the protective nosecone before the second stage separated from the third stage, and the system must function without interfering with the structure of either stage. Since the second-stage shell which surrounded the third-stage motor had to retract about five feet longitudinally without contact in order to avoid collision with the third stage, weeks of work would have to go into devising means of accomplishing that delicate feat some three hundred miles above the earth's surface. In February 1956 neither Martin nor the Laboratory dared say just how to do it. Similarly the question of whether the tankage should be aluminum or stainless steel was left unsettled. Aerojet, having had considerable experience in forming and welding stainless steel, later succeeded in persuading Martin that steel had a better strength-to-weight ratio than aluminum and would be the most satisfactory material to use. Although several Martin and NRL engineers deplored that choice, in December 1956 a backup contract with the A. O. Smith Company of Milwaukee to produce welded steel tanks would allay some anxieties.[18]

Intensive discussion of the specifications for the third stage went on concurrently with plans for the second stage. While acknowledging the necessity of lengthening the third stage and slightly increasing its diameter, Martin engineers flatly declared that the rocket could not support a thirty-inch spherical satellite unless the entire vehicle were to be bigger. Challenged by Hagen to produce mathematical proof of that assertion, they presented data on 3 February that convinced the project director. At the same time Hagen received an alternative proposal from an unexpected source, namely from James Van Allen, who was a member of the National Academy's IGY Technical Panel on the Earth Satellite Program (TPESP) and head of a newly appointed Working Group on Internal Instrumentation.

Critical of NRL's "schematic design" which, he contended, allotted only 2 pounds of the 21.5-pound body to experimenters' instruments, Van Allen urged that "half the initial group of satellites" carry a payload of cylindrical configuration, 18 inches long, 6 inches in diameter, which would reduce the weight of the inert structure from 11.5 to 5.5 pounds and thus leave 8 pounds for scientific instrumentation. This plan carried the discussion full circle. Hagen was irked. He observed that he had agreed to abandon the original, efficient conical-shaped satellite (see p. 80) to accede to the desires stressed by Whipple and endorsed without a dissenting voice by the Academy's panel. Adoption of the 20-inch sphere had been a concession that had already caused delays and otherwise avoidable expense. Furthermore, if Van Allen had counted in the telemetry equipment, he would have had to chalk up the weight allowance for instrumentation in the sphere at 10 pounds, not 2. The White House and the Defense Department had refused to act upon the Academy's plea for twelve satellite vehicles, and the Vanguard staff considered the chances slim of getting more than one successful launching out of the six authorized. An alternate design for three of the six would add to problems at this juncture. And if the Minitrack system were to fail at any point, tracking would have to depend on the visibility of the tiny object orbiting in space. The 20-inch sphere appeared to be the most practical choice.[19]

That agreed upon, the Laboratory and the Martin Company had comparatively little trouble in fixing the configuration, weight, and power requirements of the third stage. The engine was to be bottle-shaped, 55.45 to 57.5 inches long and 18 inches in diameter; the rocket's dry weight was to be about 67.5 pounds, propellant weight 395, and payload 21.5 pounds, thrust nominally 2,350 pounds in vacuum, fuel-burning time 41.5 seconds, minimum specific impulse 245 seconds, and the velocity increment 13,405 feet per second. As the government was to supply the satellite package and the structure for attaching it in and separating it from the rocket casing, the contractor had to adapt his design to fit.[20] Martin and NRL had already concluded that a dual approach to design and fabrication of the third-stage engine was desirable. So, in early March Martin placed a purchase order with the Grand Central Rocket Company of Redlands, California, while the government negotiated a direct contract with the Allegany Ballistics Laboratory of Cumberland, Maryland, to produce an alternative model.[21]

The struggle over vehicle specifications ended with the Martin Company's yielding to the Laboratory's every demand except for the thirty-inch satellite, and, in view of the strong convictions of Martin engineers that their methods and plans were generally sounder than the customer's, they yielded with good grace. NRL engineers were satisfied with the final design and indeed Milton Rosen called it "magnificent" ten years later. Many of its features were well ahead of its time. Yet the contract contained several passages—those dealing with the controversial "policy considerations"—that are likely to look deceptively innocuous to anyone unaware of their implications. Today a standard type of National Aeronautics and Space Administration contract with industry, the agreement of 1956 had to be painfully worked out to meet novel managerial problems.[22] It fixed a relationship between government and industry whereby a federal agency responsible for procuring precision hardware to use under unknown conditions of scientific exploration became manager of the project, wielding authority to direct the work of an industrial prime contractor and subcontractors. The arguments that delayed negotiations in 1955 and 1956 revolved around four issues: the parts of the system which the government declared it must procure and operate, government supervision over Martin's subcontractors, deviations from specifications, and the "margin of safety."

The final agreement listed in detail what the government was to supply, even the number of desks and chairs to be put in the blockhouse at the Cape. Appendixes I and III entitled respectively "Government Furnished Equipment" and "Government Support Facilities and Services" contained the particulars that amplified statements given in the first paragraphs of the specification document. Under the heading "General Procedures" two sentences defined the extent of NRL's authority: "The Government will exercise such direction and controls as are necessary to assure itself that launching and test vehicles meet their objectives in both performance and reliability. The Government will determine the performance and degree of reliability that can be reasonably expected within the time-scale framework." The contractor, in short, could not decide for himself what was good enough. Under the heading "Field Operations," the wording ran: "The Government will arrange and control field operations. In addition the Government will determine the requirements for and will provide equipment and services for telemetry, tracking, range safety, and data reduction. The Contractor shall supply the Government with the Contractor's requirements for data." When asked ten years later why Martin abandoned its advocacy of using telemetry equipment already available and, in company opinion, as good or better than the more complex that NRL demanded, Elliott Felt shrugged and said smilingly: "Oh, they were so insistent that it wasn't worth fighting over any longer. Besides we were tired of going to meetings at the Laboratory. It was such a dreary place." The Laboratory would have to bear the responsibility for any waste of time or money. In actuality by mid-1957 the accuracy of the NRL telemetry system would be an eye-opener to Air Force officers and contractors at the Cape.[23]

NRL also had its way in requiring the specifications for subcontractors' jobs to be incorporated in the government contract with Martin. The provision automatically put the Laboratory's staff in a position to monitor all Vanguard work in process. The technical director of the program believed it the only possible way to ensure that the quality of subsystems and parts was up to the Laboratory's exacting standards and, if not, to institute corrective measures promptly. In spite of its initial objections, the Martin Company in time to come would find government intervention helpful, for example, in flying out new heat-treating equipment to the Aerojet plant when the subcontractor was encountering troubles with the second-stage tankage.[24]

Deviation from specifications was settled by a decree that every change must be recorded in writing whether approved in conferences between the government and the primary contractor or more formally by letter. The Bureau of Aeronautics Representative, BAR/Baltimore, might sanction minor revisions; NRL must endorse all major deviations. Although Martin complained about the one-sided nature of clauses empowering the government to direct changes which the contractor thought needless or undesirable, the company knew that the customer, right or wrong, was paying the piper and could call the tune.

In the section dealing with the safety of the rocket's structure, one sentence alone took a week of angry discussion to draft in its final form: "The minimum margin of safety shall be greater than zero." Martin, relying on its experience in extrapolation, believed its calculations of stresses as reliable a guide to structural design as the state of the art could furnish. NRL, doubting the capacity of anyone to compute with sufficient exactitude the structural strengths required to meet unknown conditions, insisted on safeguards. Company officials argued that the customer must trust to company competence. And company pride was involved. What Martin considered redundancies, the Laboratory labeled necessary precautions. So one paragraph covering structural design set the minimum yield factor of safety at 1.10 and the minimum ultimate factor at 1.25. "Where structural failure would endanger personnel during ground handling, erection and checkout, . . . the minimum ultimate factor of safety shall be 1.50. . . . For pressure vessels other than fuel and oxidizer tanks, where failure would endanger personnel, . . . the minimum ultimate factor of safety shall be 2.00." These additions to the strength of the Vanguard structure, as Martin pointed out, increased the rocket's weight, but, in the view of most NRL experts, that was a lesser penalty than the loss of public endorsement of the satellite program which might well have resulted had the men working on the launcher suffered serious injuries.[25] And even if men were not hurt, the loss of material and time from structural failures would have been disastrous, for estimates made later at the firing complex showed that every hour of delay in a countdown cost at least $25,000.[26]

The wording of several passages of the specification was vague or at least subject to more than one interpretation. No one could quarrel with the goal stated under the heading "Simplicity": "Simplicity of satellite vehicle construction shall be emphasized in the interest of providing reliability and decreased weight. Every effort shall be made to keep the number and complexity of components . . . to a minimum. Applications of this principle must not jeopardize attaining the mission." The last sentence, however sensible, opened the door to new debates when testing began. Were two sets of batteries and valves, for example, a redundancy or necessary insurance? By the end of 1956 NRL acknowledged that some of its original demands were unrealistically cautious. Reliability, a universally acknowledged "must," was obviously too elusive a term to define precisely. The requirement merely read: "The vehicle shall be designed and components selected on the basis of available reliability data to insure reliability consistent with the state of the art. Reliability studies and statistical testing to establish such data shall not be required." In other words, build as reliable a vehicle as you know how to.[27]

The inspection and testing requirements, on the contrary, were explicit. The primary contractor was to conduct at his plant tests and inspections of parts, subsystems, and the assembled vehicle; inspection of materials, components, and subsystems was also to take place during manufacture in Martin's and subcontractors' shops and, after assembly of each stage of the vehicle at the Middle River plant, again under the supervision of a government inspector. Thus there was to be a triple check before firing tests began and before government acceptance of each vehicle. Procedures in Baltimore were to include, first, systems tests of stability, pressurization efficiency, and structural noise that might interfere with the functioning of the flight control system; and, second, environmental testing with simulation of vibration, shock, temperature, humidity, and pressure. But data required for design evaluation of the entire vehicle were to be obtained from flight-testing at Cape Canaveral. As the contractor was to have responsibility for directing and conducting the test programs, whereas, as noted above, the government was to "control all field operations," the seeds of future controversy were embedded in the document.

The purpose of each static firing and flight test in the field was, however, clearly set forth. The program was to progress from static and then flight testing of TV–0, the name given to the modified Viking 13 single-stage rocket, to Test Vehicle 1, the revamped Viking 14, carrying a dummy second stage and a live third stage. TV–3 would be the first live three-stage Vanguard to be fired, and TV–4 the first to carry the satellite package. The last test vehicle, TV–5, equipped with somewhat different instrumentation from that in TV–4, would also carry a satellite.[28] Only after analyzing the performance of the test vehicles were launchings of the six SLVs, satellite launch vehicles, to begin. Events in 1957 were destined to change this schedule.

One policy matter not expressly covered in the final contract was the question of releasing information to the public. Deputy Secretary of Defense

Robertson's directive of 9 September 1955 had, it is true, decreed that all releases dealing with the Pentagon's share of the program must be cleared through the Office of Security Review, and the Navy had assigned a Confidential classification to the project. But those restrictions still left areas of doubt. The Martin Company naturally wanted freedom to answer newspapermen's inquiries about company plans and progress that did not impinge on security rules; it seemed a reasonable form of free advertising. But ONR was adamant that every statement for public consumption must be cleared through government channels. The Vanguard teams at the Laboratory and at the contractor's plant strove to abide by Edward Hulburt's informal dictate that, whatever the omissions, the information must be strictly truthful. Yet well before all the design specifications were agreed upon, so many problems were arising about public relations that Hagen eventually asked Commander Berg to prepare a list of classified and non-classified items in the vehicle. A time-consuming task, it proved a valuable guide for briefings. Both Hagen and Berg believed that all Americans had a legitimate interest in the great venture. So, as Berg put it, "instead of brushing a few crumbs through the cracks to them, we invited them to the banquet table and merely omitted a few courses." [29]

The signing of the design specification in itself constituted a milestone for the program. Reached six weeks after the date originally set, the agreement cleared the road ahead; manufacturing drawings and shop work could now proceed, even though addenda and revisions to the basic document were already under discussion. The Martin Company on 8 March submitted a list of minor changes clarifying the company's responsibilities, appending a breakdown of costs now estimated at $26,212,938, and making some shifts in the test schedule. An addendum prepared by the Laboratory set forth the procedures for incorporating in the specification the decisions left unsettled in the agreement of 29 February. It also made more explicit the duties of the contractor's field crew and the relationship to obtain between Martin's field project engineer and NRL's test coordinator and project coordinator at AFMTC, and the amendment described the character of the supervision the Laboratory would exercise over subcontractors. Furthermore, it expressly assigned to the Martin Company responsibility for technical coordination of the work to be done on the third-stage rocket motor by the Allegany Ballistics Laboratory with that to be undertaken by the Grand Central Rocket Company. With these elaborations completed, both parties signed the final contract on 30 April.[30]

In the retrospect of a dozen years, the major participants viewed the battle over Vanguard specifications as inevitable, partly because similar fights, albeit on a smaller scale, still occur when a much-publicized contract is at stake, and more largely because the 1955–6 struggle, involving as

94

it did a controversial principle of management, had to go on under time pressures so severe as to make tempers peculiarly edgy. Just as the differences of opinion over the design, as such, rarely if ever created animus, so by 1967 the former antagonists could see that neither side had been wholly and consistently right. Under the circumstances obtaining in the mid-1950s, it was little short of a miracle that a workable modus operandi had come into being by spring 1956.

6

ROLE OF NAS AND TPESP, 1955-1956

IN AUTUMN 1955, while the Naval Research Laboratory and the Martin Company were starting work on specifications for the vehicle, the United States National Committee for the IGY had organized a Technical Panel on the Earth Satellite Program (TPESP) to watch over the purely scientific phases of the project, to select the experiments to install in the birds, and, subject to USNC and Executive Committee approval, to fix the policies and procedures in regard to financial commitments, institutional relationships, and educational releases to the public. In assessing the work of the National Committee, the technical panels, and the working groups—under the National Academy of Sciences (NAS)—it is essential to realize that, with the exception of the seven to eight men who composed the USNC's secretariat and the handful of scientists whose government jobs extended to participation in IGY planning and action, all members gave their services without compensation. Academy funds, derived from the National Science Foundation, paid travel expenses and in some instances a modest per diem to cover living costs during protracted committee and panel sessions. But most of the men who spent days, sometimes weeks, in directing IGY programs volunteered their time, despite demanding professional obligations on university campuses or in industry. In the earth satellite project, moreover, panel and working group members also risked their future standing in the scientific world, for the success of the venture was far from assured and failure could end in the men responsible for the scientific aspects of the undertaking being labeled incompetent.

The Technical Panel on the Earth Satellite Program consisted of the chairman, Richard Porter, who was also a member of the Stewart Committee, Joseph E. Kaplan and Hugh Odishaw, USNC chairman and secretary respectively, Homer E. Newell, Jr., of NRL, William H. Pickering of the

Jet Propulsion Laboratory at the California Institute of Technology, Athelstan Spilhaus of the University of Minnesota, Lyman Spitzer, Jr., of Princeton University, James A. Van Allen of the State University of Iowa, and Fred Whipple of the Smithsonian Astrophysical Laboratory. After Spitzer resigned in 1956, Gerald Clemence of the Naval Observatory and Michael Ference, Jr., of the Ford Motor Company brought the number to ten, and in the months after Sputnik the addition of Alan Shapley of the National Bureau of Standards and W. W. Kellogg of the RAND Corporation further enlarged the membership. By invitation, a few guests and observers always attended the meetings to contribute information or represent such agencies as the National Science Foundation, the DoD Comptroller's Office, and the office of the Assistant Secretary of Defense for Research and Development.[1]

The most urgent task of the panel when it first met was the drafting of a budget, since if the USNC was to get adequate supplemental funds in fiscal year 1956, the request had to reach the Science Foundation in the near future and go thence to the Bureau of the Budget and Congress. The panel's main quandary was what to expect from the Department of Defense in keeping with the Department's pledge "to furnish logistic support for the NAS–USNC Satellite Program within reasonable limits." What were reasonable limits? The commitment as it stood encompassed plans for "six earnest tries" at a satellite launching, but the Academy's goal was twelve, since scientists, doubtful of being able to locate in space and track every bird put into orbit, feared that fewer than twelve attempts would imperil acquisition of the scientific data at which the program was aimed. Irrespective of how many of the six shots authorized to date were successful, would the Defense Department share the cost of an extended program? The panel decided to divide its budget estimates into two phases, one covering from April 1956 through March 1958, during which time the six tries already approved would presumably be completed, and Phase II running for another one and a quarter years; assuming continued DoD participation, Phase II should permit another six launchings. The panel appointed a three-man working group to prepare this double set of estimates.[2]

When the working group tendered its suggested budget in November, the report set $8.337 million as the sum needed to cover National Academy expenses for instrumentation for the six operation vehicles, ground stations, and "certain scientific personnel" during Phase I; a "guesstimate" put Defense Department costs at about $19.1 million. Optimistically a staff member of the USNC secretariat explained that while "areas of overlap of responsibilities" between the Academy and the military would doubtless occur, "it is expected that mutual understanding will prevent difficulties."

Inasmuch as the Defense Department had not yet agreed to support Phase II, figures for extension of the program through June 1959 were more tentative, but an additional $20 million appeared to be enough. Of that amount about $7.1 would be Defense money; IGY funds, if need be, could logically meet the $6 million cost of procuring six additional vehicles. All told, NAS–NSF expenditures for Phase II would come to approximately $10.87 million, a figure that would keep Academy expenditures for three and a quarter years well below the $20 million ceiling set by the National Committee for the satellite program. "With the understanding that some latitude would be given for minor adjustments," the panel approved the estimates.[3] The costs to be financed by the Science Foundation are shown in the table (p. 100).[4]

In view of the stress the Stewart Committee's original report had laid upon the advantages of the Minitrack system, the nearly $3 million allotted in the IGY budget for optical tracking may seem surprising. Years later John Hagen remarked that, as things turned out, the program would have lost little from omitting provision for optical tracking. But in 1955, and indeed long afterward, most of the IGY committee and satellite panel believed tracking by cameras, telescopes, and theodolites vitally important. Not only would it be more accurate, but it might well be a more dependable, albeit a more restricted, method than a still untried electronic system. "One should act," ran the panel minutes, "as though there were only a 50–50 chance of the Minitrack's operating successfully (although actually the Minitrack engineers estimate that on any one firing the chance of successful operation should be better than 95%)." Athelstan Spilhaus furthermore contended that many people would have little faith in information relayed by telemetry, that form of occult magic; they would put stock in a man-made moon only if they could see it with the help of simple instruments as it passed overhead. Consequently plans for optical tracking received careful attention. To begin with, a Working Group on Optical Observations and Tracking consisting of Fred Whipple and Lyman Spitzer presented a scheme of enlisting the help of professional astronomers, training teams of amateur observers, and setting up central administrative, computing, and analysis facilities. The panel promptly voted to ask the USNC to obtain from the Science Foundation a grant of $50,000 for the Smithsonian Astrophysical Observatory in Cambridge, Massachussetts, to oversee the establishment of twelve observation stations, arrange for procurement of special equipment, secure the collaboration of governmental and other professional scientific groups, and recruit amateurs for what later came to be called "Moonwatch." The plan of operation appeared in the diagrammatic form shown on page 101.[5]

	Launching & Propulsion	Satellite Instrumentation	Radio Tracking & Telemetry	Optical Tracking	Orbital Computations & Data	Scientific Coordinates	Total
Salaries	$ 53,000	–	$1,365,000	$ 936,000	$ 279,500	$ 90,000	$ 2,723,500
Travel	3,000	–	273,000	240,000	30,000	26,000	572,000
Transport of Things	5,000	–	360,000	84,000	5,000	–	454,000
Communications	2,000	–	54,000	49,000	19,500	5,000	129,500
Supplies and Materials	31,000	–	450,000	393,000	64,000	8,000	946,000
Equipment and Facilities	6,090,000	1,410,000	4,834,000	1,296,000	804,000	3,000	14,437,000
Total	$6,184,000	$1,410,000	$7,336,000	$2,998,000	$1,202,000	$132,000	$19,262,000

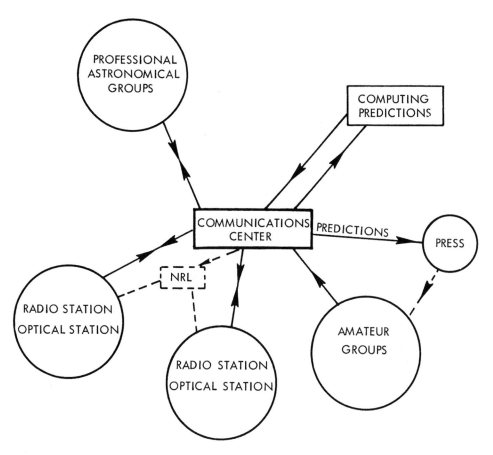

Diagram of the plan of operation for Project Moonwatch.

Although formal action on the plan had to wait upon passage of a supplemental appropriation for the IGY, Whipple at once set about organizing SAO's optical tracking scheme.

The enthusiasm of amateur star-gazers over the opportunity to have a part in an important scientific venture early impressed the USNC's Executive Committee. Here evidently was a simple way of widening public interest in the IGY both at home and abroad. In response to the committee's suggestions that the men in charge of other phases of the IGY open the doors to amateur participation, in early 1956 the technical panel authorized "Moonbeam" for ham radio operators; the Naval Research Laboratory accepted responsibility for indoctrinating licensed applicants in the essential do's and don't's, chiefly by means of a descriptive and technical manual to be prepared by John Hagen. At the same time Whipple agreed to draft instructions for Moonwatch teams.[8]

When hearings on the IGY supplemental budget began on the Hill in March 1956, spokesmen for the Science Foundation and the National Academy had some difficulty in justifying the request for $19.262 million for the satellite program, by far the largest single item in the $28 million budget. Patiently they explained that the undertaking had assumed new dimensions since the White House announcement of July 1955, that more nations, notably the U.S.S.R., were now participating, and that American universities would bear much of the expense of interpreting the scientific data picked up from the satellites. Congressional questions hammered away at why the Academy planned to share its findings with other nations and yet was asking the United States government to spend millions of taxpayers' money to acquire the information. Would anything useful, congressmen asked, come out of the program? To a question posed by a senator, "How do you propose to get results from devices in the satellite?" Joseph Kaplan replied—

> We literally talk to the satellite, using radio waves, which is a technique known as telemetering . . . .
> In conventional rocket research, the techniques of recovering equipment by means of parachutes has been well worked out . . . . But in the satellite nothing but the radio signals will be recovered at first.[7]

Richard Porter defended the request for money enough to expand the number of launchings from six to twelve: twelve, in the judgment of the Academy's technical panel, represented "a realistic program." Although Congress in the end pared $1 million from the Foundation's IGY budget and NSF and the USNC then reduced the total for the satellite program to $18.364 million, the panel believed that it could still manage to finance the twelve shots. Kaplan's summary of events ran: "Berkner sold the program to the Academy; Waterman sold it to the White House; I sold it to Congress." He did not add that nobody sold it to Secretary of Defense Wilson or that Congress, though granting funds to the Science Foundation for it, never appropriated a cent to the Navy for the vehicle.[8]

Meanwhile the TPESP had set in motion a plan for obtaining proposals for scientific apparatus to put into the instrumented satellites. Recognizing that selection of the most rewarding experiments would be simplified by full discussion of possibilities among geophysicists, the TPESP decided to organize a symposium on "The Scientific Aspects of Earth Satellites." By briefing participants on the limitations which the vehicle would impose on satellite instrumentation, the panel should avoid exposure to a flood of impractical ideas and find the choice among useful proposals easier. The originators of the approved experiments would then receive grants from the Science Foundation to assist them in developing the implementing instruments. Publication of abstracts of the papers presented, furthermore, would serve the purpose of informing foreign scientists about the scope of the American program. Van Allen and Odishaw to whom the panel entrusted the task of arranging the symposium concluded that the tenth anniversary meeting of the Upper Atmosphere Rocket Research Panel scheduled for late January 1956 would be the most suitable occasion, for it would bring together men already familiar with high atmospheric research.

Van Allen's letter to members of the rocket group explained that the symposium was to be kept to not more than fifty persons, a number that would embrace most American scientists qualified by training and experience to contribute to a meaningful discussion in this little known field. The papers to be read should be "specific, critically considered, and pertinent to the present state of technology and the reasonable projection of this technology into the next few years." Plans for physical experiments and observations, theoretical and interpretative matters, and techniques and components of a novel nature would all be acceptable topics, but not space medicine, or the legal and political aspects of the satellite program, or essays dealing with vehicle propulsion and guidance.[9]

Although the scientists who presented papers two months later at the Ann Arbor symposium generally adhered to the rules Van Allen had laid down, the problem of appraising the relative value of their propositions led the panel to ask Van Allen to head a Working Group on Internal Instrumentation to screen out the best ideas for later panel consideration, unobtrusively to elicit proposals in neglected fields, and to recommend priorities for "on-board" experiments to be flown in the first satellites.[10] To help the panel reach decisions on other complicated technical problems, in March the TPESP created also a Working Group on Tracking and Computation chaired by Pickering. Inasmuch as the panel had to review working group recommendations carefully and, where they involved IGY finances and international relationships, had to submit them to the USNC and its Executive Committee for final endorsement, the arrangement meant a good deal of thrashing over the same ground. But no other method of handling IGY business was feasible. Academic scientists, upon whose knowledge the usefulness of the IGY depended, had to parcel out their IGY chores among each other and rely upon reports to keep themselves abreast of what their fellows were doing. That the system worked at all was due chiefly to Richard Porter's contributions of time and thought and to the efficiency of Hugh Odishaw and his assistants on the USNC secretariat.

During the spring of 1956 the panel discussed a variety of questions: how, for example, to evaluate the chances of a successful Russian satellite launching before the year was over, how to explain to the public the goals of the American program effectively, and how to obtain from the Naval Research Laboratory greater flexibility in satellite design and the telemetry and tracking systems to be employed. Discreet inquiries through the ruling international body for the IGY, CSAGI, appeared to be the best way to verify or put to rest the vague rumors of an early Russian satellite launching. Speakers at international scientific gatherings scheduled for the summer and autumn and at meetings of American societies should be able to disseminate accurate information about the Vanguard project. Publication in May of the abstracts of the papers read at Ann Arbor offered an impressive summary of what the program might accomplish. NRL, however, stood by its contention that all experiments installed in the birds should rely on the Minitrack and the types of telemetry NRL had already chosen. As for the satellite, Homer Newell reported that the Laboratory was going to produce at first some 6.4-inch as well as 20-inch spheres and thereafter would try other configurations, including perhaps the cylinder-shape that Van Allen advocated. Heavy instrumentation to check the performance of each stage of the launcher had to go into the test vehicles, but the Vanguard staff concluded that every test vehicle containing three live stages could carry also a small, lightweight satellite. If the 6.4-inch, 4-pound bird got into orbit, that success would not only testify to the adequacy of the launcher but permit later flights, unburdened with the test instrumentation, to carry the bigger spheres and more scientific paraphernalia.[11] All these problems, however, faded into the background when the panel discovered that a budgetary crisis was putting the entire satellite program in jeopardy.

Paul A. Smith of the Department of the Navy had warned the satellite panel in March that then estimates pointed to a possible $20 million deficit in DoD funding for Vanguard. But panel members had not been seriously alarmed. Academy scientists, the Vanguard comptroller at NRL ruefully observed, looked upon the DoD as a "fat cat" that could afford anything it wanted. Surely the Department would find some way of honoring its commitments even though they were greater than originally expected. And, with over $18 million appropriated to the Academy for the satellite program, the USNC could pay for the second six vehicles and thus carry out Phase II. In April 1956 Newell and Ross Peavy of the USNC secretariat figured that the distribution of expenditures between the Academy and the Defense Department would be approximately as follows: [12]

	Total USNC Phase I	USNC Phase II	DoD Phase I
1. Launching Vehicles	0	$ 6,094,000	$37,200,000
2. Satellite Instrumentation	$ 610,000	610,000	0
2a. Scientific Data Telemetering	1,362,000	382,000	0
3. Radio Tracking	3,796,000	2,076,000	2,400,000
4. Vehicle Telemetering (launching phase)	(included in item 1)		2,900,000
5. Optical Tracking	2,233,000	765,000	0
6. Computation	251,000	951,000	1,300,000
7. Scientific Coordination	82,000	50,000	0
Totals	$8,334,000	$10,928,000	$43,800,000

A blow awaited the scientists when they met in June. A statement from the Pentagon informed them that since the preceding October the estimated costs to the military of providing six launching vehicles, building tracking stations and furnishing logistic support had nearly trebled and now ran to some $63 million. Only $21 million had been authorized. Secretary of Defense Wilson had consequently outlined to President Eisenhower four possible courses of action: (1) proceeding with the full NAS proposal of six pilot launchers and then six more; (2) proceeding with the six launchings already approved; (3) proceeding to the extent of available funds; or (4) canceling the program. Upon the advice of the Defense Department and the National Security Council, the President had chosen course 2 and was requesting the Bureau of the Budget to find additional funds for it. Worse followed: the emergency fund on which the Naval Research Laboratory had been drawing to pay contractors' fees was virtually exhausted, and the Defense Department, knowing that the congressional appropriation for the IGY included $10.728 million for a second six satellite launchings, now asked the Science Foundation and the Academy to turn over to the Navy $6 million immediately to prevent work stoppages on the test vehicles.

Feeling that the pistol had been put to its head, the TPESP reluctantly admitted that it had little choice. It would have to acquiesce in the transfer of the $5.8 million that would be surplus in the IGY budget if only six firings were to occur. The President had decreed only six earnest tries. But panel members insisted that a letter to the Science Foundation make clear that they would accept the reduction in the program and the sacrifice of IGY money only under duress and not by free will. They would continue to seek means of restoring the number of launchings to twelve, since half that number would probably mean at best getting not more than two satellites into orbit. They drew some consolation from a message from Alan Waterman "describing his understanding that 'the need for and feasibility of constructing and launching six additional satellites will be a subject of review on the part of DoD.' " With that bit of reassurance in mind, the scientists stipulated that release of their surplus funds to the Defense Department "should not preclude the ultimate availability of equivalent funds if the expanded program should be deemed again feasible." The USNC and the Executive Committee agreed to the transfer of $5.8 million a week later.[13]

The indignation of the IGY group subsided somewhat during the summer. That the threatened work stoppage on Vanguard did not occur lightened the earlier tensions considerably. Over dinner at the Cosmos Club in mid-July Porter, Waterman, Admiral Rawson Bennett, Director of ONR, and Clifford Furnas, who had succeeded Quarles as assistant Secretary of Defense for Research and Development, had "a very useful and constructive discussion," as Porter described it. Over the division of responsibilities between the Academy and the military services, a differentiation supposedly clearly defined in 1955, confusion had risen in the intervening months. The four distinguished men, each speaking for his own organization, reached a general understanding that, in Porter's view, ironed out many difficulties. In a letter to Kaplan, Porter summarized the results:

> 1. The Earth-satellite program should be thought of as an IGY project in which the Department of Defense is cooperating, rather than as a D.O.D. project.
> 2. The D.O.D. has accepted the responsibility for development, procurement, and launching of the vehicles, and for demonstrating that the vehicles have or have not successfully been placed in orbit, and for providing certain other technical and logistic support.
> 3. The National Academy of Sciences retains the over-all responsibility for all phases of the scientific utilization of the satellite vehicles including precise determination of orbit and other measurements which may be made on . . . the vehicles for scientific purposes. Funds for this work have been appropriated by the Congress to the National Science Foundation, and the Foundation must properly account for these funds.

Later paragraphs underscored the Academy's delegation of authority to NRL to develop, procure, and operate high-precision radio tracking and telemetry stations, perform the analysis and computations of data, and devise and construct "certain instrumented experiments" to be carried on board the satellite. For those phases of the program, Porter's interpretation thus made the Laboratory an agent of the Academy. The Laboratory was to report its fiscal estimates and commitments in these areas regularly to the Academy and the Foundation. Publicity releases must obtain security clearance from the DoD and "policy" clearance from the IGY committee after consultation with the State Department. The USNC and the Defense Department would issue an official "Policies and Responsibilities document" as soon as both bodies had agreed upon the exact wording.[14]

The TPESP rested easier upon learning that Furnas, Bennett, and Waterman recognized the satellite program as an Academy project, in no sense a military venture. To the USNC, with its responsibilities to the international scientific community, it was important to have all doubts on that score removed. The verbal agreement, committed to paper only in Porter's personal letter to the USNC chairman, could not, however, change the sober realities of the situation. The major tasks of the satellite undertaking still fell to NRL, to the men who had originated the Vanguard design and its tracking and telemetry systems, and to those responsible for getting the satellite into orbit. The Vanguard team at the Laboratory had too much respect for the stature of the scientists on the IGY panel to dismiss their ideas lightly, but the men sweating over the job of producing a reliable launcher and dependable tracking and radio communication did not welcome gratuitous advice on how to do it.[15] Although labeled an Academy program, in actuality its execution was divided into two unequal parts, the most costly and technically difficult of which was under Defense Department aegis. It was the problems arising out of this division of responsibility that later brought the National Aeronautics and Space Administration into the picture.

If few people as yet fully perceived the handicaps of split control, in December 1956—when, at the USNC's request, an ad hoc group consisting of Porter, Van Allen, Lyman Spitzer, William W. Kellogg of RAND, Newell, and Milton Rosen of NRL drafted a statement of the fields of research that a long-term, post-IGY satellite program should encompass—the group volunteered its comments on organization:

> for an extended scientific program of national scope, . . . it is important that *clear civilian authority* (as by the National Science Foundation) be established for the planning and execution . . . . In particular, it seems important to establish at the very beginning . . . *a single comprehensive budget which will include all expenditures in connection with the program, including those to be made by organizations within the military establishments.*[16] [Italics added.]

Before the autumn of 1958, the Department of Defense was accountable for expenditures for the satellite vehicles and radio tracking, the Science Foundation for the rest. Theoretically everyone understood these compelling facts. The TPESP, without relinquishing its plans for twelve launchings, had to acquiesce not only in the transfer of $5.8 million to the Navy in summer 1956 but also in the release of another $5.5 million of IGY money in October. The result was to perpetuate strains in the relations between the Science Foundation and the Academy and to encourage the panel to watch the Laboratory's performance with an increasingly critical eye. Homer Newell fortunately was usually able to explain to his panel members why engineering considerations obliged NRL to adopt features for the satellite that were unwelcome to the scientists and why delays occurred in the test program. Yet at times administrators of the Science Foundation felt impelled to intervene to keep relations between the Academy and the DoD on an even keel. Waterman, who had served as chief scientist at the Office of Naval Research before he became director of the Foundation, was a close friend of Admiral Bennett and at the same time understood the point of view of the gifted men attached to the Academy. Quiet and unassuming himself, he was an indispensable go-between in what was often a trying situation. His role was made harder, however, by the fact that most members of the TPESP and its working groups could allot only part of their time to the satellite program. They were only intermittently in Washington, and correspondence was a poor substitute for direct contact with day-to-day developments.[17] Anxiety over progress on the launch vehicle crept into panel discussions increasingly often.

Relatively little controversy arose over NRL's plans for tracking, orbital computations, and data reduction. While the panel inclined to think Army estimates for constructing and operating the Minitrack stations excessively high, the TPESP swallowed its protests, inasmuch as Defense money was to pay the costs. The Working Group on Tracking and Computation, which scrutinized every feature of NRL's Minitrack and projected computing system, questioned every detail but generally endorsed the Laboratory's arrangements. A NRL contract with International Business Machines was especially satisfactory, partly because of the free services the company offered the government. Despite this generous arrangement, the prospective costs of data reduction were beginning to rise alarmingly.[18] A contract with Radiation, Inc., of Melbourne, Florida, fixed a price of $458,000 solely for equipment and services for data reduction in connection with the telemetry to be used in launching the vehicle.

By comparison, the Smithsonian's optical tracking scheme looked cheap and easy, although scientists eager to obtain continuously recorded and telemetered data from instrumented satellites may well have regarded it as an extravagance at any price. The establishment and operation of twelve observing stations would cost nearly $2.371 million, but other expenses would be fairly light. IBM had offered free access to the 704 computer at the Massachusetts Institute of Technology for an hour a day through June 1959 and supporting services elsewhere for orbit calculations. The Army Signal Corps was willing to release a number of elbow telescopes for amateur use at military installations and the Air Force Aeronautical Chart and Information Center was ready to supply without charge some manpower and equipment for optical observation stations on islands in the Pacific. The panel, however, felt obliged to stipulate that all operators at these stations must be civilians to quiet fears of other nations that the United States was using the IGY for military ends. Yet the proffer of free services notwithstanding, as matters stood in the fall of 1956, panel chairman Porter envisaged a deficit of $1 million for the scientific phases of the satellite program.[19]

International cooperation was a basic element in the conception of the IGY, but whether other countries could or would contribute much to the American satellite undertaking long looked doubtful. Although scientists attending a western hemisphere regional conference of CSAGI in Rio de Janiero in July 1956 listened attentively to lectures given by John Hagen and William Pickering on the technical aspects of the American project, Gilman Reid of the USNC's secretariat reported: "Apparently very little information on the satellite had reached the South Americans and they regarded the program as exclusively a United States effort. In the Working Group Session . . . there was small attendance and an apparent lack of interest on how there might be participation in the program." This apathy was disconcerting, since several tracking stations along the north-south "fence" were to be located in South America. The International Committee, CSAGI, promptly urged that every National Committee appoint a "satellite reporter" to facilitate the channeling of information.[20] Lloyd Berkner, American representative on CSAGI, already was functioning in that capacity for the United States. CSAGI made further constructive recommendations at the meetings in Barcelona in September: every National Committeee should report on its capacity to set up amateur observation stations, and standardization should everywhere obtain in radio tracking and telemetry systems so as to correspond to those announced by the United States. Members of the American delegation pushed the theme of collaboration with "strong invitations" to other countries to participate; the United States National Committee would issue a technical manual to assist them.

At this point the Soviet National Committee formally announced to the international gathering that the U.S.S.R. also had inaugurated a satellite program, "by means of which measurements of atmospheric pressure and temperature, as well as observations of cosmic rays, micrometeorites, the geomagnetic field, and solar radiation will be conducted. The preparations for launching the satellite are presently being made." Although by no means certain about how much technical information the Soviets would release, Americans were relieved to have Soviet intentions brought into the open. Hugh Odishaw suggested to the panel that autumn that here might be the means of inducing the White House to restore the Academy's twelve-vehicle program. An inquiry sent a month later from the USNC to the Russians asked advice on where the United States should set up observation stations to track future U.S.S.R. satellites. As half expected, no explicit answer was forthcoming.[21]

Publicizing of satellite plans at the CSAGI sessions nevertheless benefited the American program. Australian scientists opened negotiations for a Minitrack station at Woomera which they would maintain and operate. England, France, and Italy expressed readiness to contribute various services. And the World Meteorological Organization was considering offering its network for the transmission of data.[22] Whipple reported that prospects were good for recruiting Moonwatch teams in Australia, Japan, South Africa, and possibly India, and that, provided the United States supplied the cameras, Australia, Spain, Ethiopia, Iran, and Japan wanted to establish optical tracking units. All told, by late autumn 1956, months before Lloyd Berkner and members of the secretariat completed the 415-page handbook entitled *Rockets and Satellites,* the response of the international scientific community was gratifying. It would not pay the bills, but it should inspire the White House and the American Congress to support the venture on a generous scale.

In any case the Academy must keep other nations informed about American plans and progress. The satellite manual went out in draft form to all national committees in August 1957, an amended version in November after the first Sputnik flight. Every National Committee, the TPESP agreed, should receive advance notice of launching schedules, the predicted orbit of each bird, what telemetry signals to watch for in tracking, the nature of each onboard experiment, and the method of coding data. Within five months of a successful launching, standard astronomy periodicals with an international circulation should carry detailed reports on orbital observations, and the results of scientific experiments should be released within eight months "in reduced, corrected and calibrated form," together with pertinent interpretations.

The panel's chief worry as 1957 approached was the steadily rising cost of every phase of the scientific program, particularly the Baker-Nunn tracking cameras, orbital computation and data analysis, and the sums needed to supplement the grants to scientists preparing apparatus for onboard experiments. Porter and Van Allen argued that if expenses must be cut sharply, it would be better to drop one or more tracking stations

110

rather than to reduce the funds for experiments. Meanwhile the TPESP must make a final decision about which experiments were to be flown in the first satellites.[23] Postponed until Van Allen and his Working Group on Internal Instrumentation had analyzed the pros and cons of every proposal, the selection could not be put off further.

7

ONBOARD EXPERIMENTS AND INSTRUMENTATION

THE FIRST plan for an onboard satellite experiment to reach the chairman of the USNC was "A Proposal for Cosmic Ray Observations in Earth Satellites" submitted by James Van Allen, George Ludwig, and several colleagues at the State University of Iowa. It bore the date 28 September 1955, several days before the National Committee appointed the technical panel. Eager to start on the scientific work that constituted the reason for setting up a satellite program, the Iowa group had not waited for information on the configuration of the Vanguard satellite or the weight of instrumentation it could accommodate, or whether it could be launched into either a polar or an equatorial orbit. Van Allen's covering letter set forth his belief "that the needs and desires of those contemplating use of the vehicles for scientific work [should] be adequately taken into account in connection with all major technical decisions." He took for granted that "a technical committee of broad interest and competence" would decide on "the assignment of payload space" in each satellite. As he assumed also that the program would be "a continuing one in which there will be many satellites flown over an indefinitely extended period of time," he envisaged development of a succession of vehicles, capable of placing in orbit increasingly capacious satellites from two feet to three feet and over in diameter and weighing from five pounds to fifty and more.[1]

It is not surprising that Van Allen had an experiment planned so promptly, for he had had long talks with Ernst Stuhlinger, chief scientist at the Army Ballistic Missile unit, when both men were at Princeton in 1953–1954. Stuhlinger's technical knowledge and vision had inspired the younger man to prepare for the day when satellites would carry scientific instruments beyond the earth's atmosphere. What he learned from Stuhlinger about the work afoot at Huntsville, furthermore, led him to believe in 1955 that a sizable payload would be possible in an IGY satellite, even though he realized that the Vanguard first stage would have far less power than the Redstone rocket. In fact, as he acknowledged long afterward, from the beginning he designed the pot of instruments for the cosmic ray observations in a form that would readily adapt it to installation in the satellite of the bigger vehicle were that eventually to be available.[2]

"Cosmic ray observations above 50 kilometers altitude," Van Allen stated in the original proposal, "have a special simplicity and importance because only above such altitudes can one's apparatus be placed in direct contact with the primary radiation before its profound moderation in the earth's atmosphere." Over the preceding nine years he and his associates had pursued investigations by means of sounding rockets, but an instrumented earth satellite could provide in a week more satisfactory data than scientists could obtain from rockets in twelve years of work. A worldwide survey from a satellite would furnish information on the geographical distribution of arriving cosmic radiation and permit deductions about the magnitude and nature of what is solar in origin. If the satellite orbit were pole to pole, or even equatorial, the survey would produce a mapping of the earth's effective geomagnetic field and reveal the correlation of fluctuations in cosmic ray intensity with terrestrial magnetic and solar activity. Cerenkov detectors, already successfully used in Skyhook balloon flights, could measure the relative abundance of the light elements of cosmic radiation—hydrogen, helium, lithium, beryllium, boron, carbon, nitrogen, oxygen and fluorine—and thus establish the distribution of nuclear species in the primary radiation before it encounters the atmosphere. Simple measurements would also increase knowledge of the nature of the cosmic ray albedo of the atmosphere. (Albedo is the ratio of the amount of electromagnetic radiation reflected by a body to the amount incident upon it.) As the albedo consists of products of nuclear reactions, in the upper levels of the atmosphere, which happen to proceed in upward directions, measurement of the total cosmic ray intensity as a function of distance from the earth should permit determination of the magnitude of the albedo. A simple detector in the satellite should, moreover, furnish means of charting the arrival of auroral radiations at the top of the earth's atmosphere.

The necessary instrumentation, consisting of a Geiger counter, Cerenkov detectors, and telemetry equipment using conventional batteries would not weigh more than fifty pounds, or, if solar batteries were available, not more than thirty pounds. If the telemetering of data were to be continuous to ground stations, the usual methods of data transmission would be used; if communication were to be intermittent, a coded integrated system would be employed. Preparatory work at the university laboratory could begin at once, and the flight apparatus could be completed in about a year, provided the government-supplied telemetry system were ready by then. The experimenters wanted to share in choosing the characteristics of the system and in its preliminary testing. They planned to construct twenty sets of apparatus, fifteen to expend in preliminary tests, five to be flown in satellites. Van Allen, whom a government financial expert once characterized as "a small-town boy, a backyard scientist" who believed in keeping things simple, estimated the cost for approximately three years' work at $66,125, including $1,000 for the time spent on reduction of the scientific data and publication of the findings. That estimated cost would rise six months later to $106,375.[3]

A second proposal came a few weeks later, when Fred Singer of the University of Maryland outlined a plan for "Measurement of Meteoric Dust Erosion of the Satellite Skin." Singer's idea was to design a radioactive gauge to place on the satellite's shell in order to measure the flux of integrated cosmic dust and compare the results with those obtained by optical observation from the earth. The method should gauge the effects of erosion on different surfaces of the satellite and reveal the changes of surface and aerodynamic properties as well as the subsidiary effects on satellite albedo and temperature. He proposed to measure erosion by observing the decrease in the activity of a radioactive portion of the satellite skin. A beta ray detector could monitor the activity on the interior surface of the shell by incorporating in the skin such beta emitters as phosphorous 32, strontium 89, and others. The investigator could analyze the resulting data in comparison with those obtained in the laboratory from charging dust particles, accelerating them electromagnetically, and then examining the surface under a microscope. The experiment would take about two and a half years to complete and would cost about $52,900.[4]

Because of Hagen's reluctant agreement to discard plans for a conical satellite and employ a sphere instead, it was late November before Homer Newell could present to the TPESP a summary of what the major characteristics of the Vanguard satellite were to be. Since the entire satellite was to weigh in the neighborhood of ten kilograms, Newell explained, only about one kilogram could be devoted to the scientific payload exclusive of the telemetry and batteries. That news was obviously disconcerting to Van Allen. The satellite system was to include two concentric spheres, an outer sphere twenty to thirty inches in diameter, and a smaller central sphere, about a foot in diameter, which would house much of the research instrumentation. Each sphere would be pressurized independently with helium. Welded seals were to be used throughout. "The various equipments will, however, be designed so as to operate even though the outer sphere loses its pressure." Just as tracking considerations dictated the configuration of the shell, so the Laboratory designers believed a spherical inner container would simplify temperature control.[5]

"Temperatures between 5° and 50°C will be acceptable to all of the items operating in the satellite." Transistor characteristics were the principal limiting factor. "More advanced transistors probably will be able to operate successfully in the range from −20° to +80°C. It is expected that this temperature range can be maintained within the central sphere provided the surface sphere is coated with an appropriate material such as ALSAC. This material is highly reflective to solar radiation, yet highly emissive with respect to infra-red radiation."

"Spin rates between 250 and 400 r.p.m. will probably occur in the Project Vanguard satellites." Expectations ran that all equipment could be so designed as to withstand the accelerations to which the vehicle would subject it. The telemetering system was to be tied in with the Minitrack system. Estimates put the telemetering reception interval during each revolution at a minimum of eight seconds at a 200-mile perigee, but an increase to as much as a minute might be possible.[6]

Every scientist preparing an experiment would have to take these conditions into account. Transistorized circuitry, desirable because of its light weight, had the disadvantage of sensitivity to extremes of temperature. Similarly the capacity of the satellite's outer shell to resist puncture by meteoric and micrometeoric particles and to withstand the action of atmospheric ions would affect the level of pressures on the inner sphere and might thus modify performance of the instruments housed in a central insular region. To accumulate exact data on surface and internal temperatures, surface erosion, and internal pressures, NRL proposed to conduct environmental studies in the first satellites launched.

Hermann LaGow and several NRL associates were designing instrumentation for these studies. A pair of thermistors mounted on the satellite's outer surface and one thermistor on the "instrumentation island" thermally insulated from the skin would measure the temperatures. Surface temperatures would probably vary widely as the satellite passed successively through daytime and night conditions, but changes would probably be relatively small on the instrument package itself. To gauge pressures inside the outer sphere NRL planned to install snap switches which would relay signals when the internal pressures dropped below predetermined levels; a final signal would start the operation of a Pirani gauge to measure leakage rates. The experts believed it possible to distinguish between leakage caused by meteoric punctures and that caused by imperfections in welding and sealing of the outer sphere. A coating, with suitable electrical resistance characteristics, on a portion of the satellite's outer skin was to form a circuit element capable of transmitting data on the rates of surface erosion; as erosion decreased surface thickness, resistance would increase. About five seconds would be enough telemetry time to transmit the data to ground stations. The instrumentation would weigh about one hundred grams, the miniaturized batteries about two hundred.[7]

At the same time Newell described two physical experiments that NRL considered well adapted to installation in an early Vanguard. Either of these could be flown along with the equipment for the environmental studies and use the same batteries. Leslie H. Meredith had prepared one of these proposals, namely, an investigation of the rigidity spectrum of primary cosmic rays, but he withdrew it some months later because it appeared unlikely to net enough information to be worth pursuing. The other, more promising, scheme came from Herbert Friedman and associates in the Electron Optics Branch of the Laboratory's Optics Division. They sought to determine the variation in the intensity of solar Lyman-alpha radiation during each revolution of the satellite about the earth. Lyman-alpha radiation is the emission from the strongest line within the ultraviolet region of the hydrogen atom's spectrum. Light of this short wavelength is not transmitted by the earth's atmosphere, but delicate instruments in a satellite might record the increases in the radiational intensity to be expected during the minutes of a satellite's flight through sunlight in comparison with the level of radiation registered during darkness. The experimenters planned to use an ion chamber sensitive only to the narrow region of the spectrum centered on the Lyman-alpha line, and circuitry to store the peak signal developed by the detector. Photoionization of the nitric oxide filling the ion chamber would create the spectral sensitivity. A photocell would relay data on the satellite's aspect relative to the sun. Five seconds would probably suffice to read out all information by telemetry. The instrumentation, capable of operating continuously for approximately five hundred hours, would weigh about six hundred grams and occupy about five hundred cubic centimeters of space in the instrument package.[8]

Thus at the end of 1955, a month before the creation of the Working Group on Internal Instrumentation, the TPESP had on hand five possible experiments to consider. As anticipated, the symposium held in Ann Arbor late in January 1956 brought in an additional crop of proposals—indeed, a number of sufficient interest to the scientific world to warrant publication in book form.[9] But by no means all the presentations described specific experiments: some dealt with problems awaiting solution but offered no explicit plan for using instrumented satellites to answer the questions; a few papers were directed at engineering or tracking techniques; one or two called for the use of apparatus which admittedly was unlikely to be perfected during the IGY. Nevertheless, by March, when the panel's Working Group on Internal Instrumentation held its first meeting, the WGII, so-called, had before it eleven propositions that merited serious consideration and another four that needed clarification.[10]

Of the men whom Van Allen chose to serve with him, Porter, Odishaw, and Lyman Spitzer, as members of the parent panel, and Herbert Friedman of NRL were already familiar with the satellite program; William W. Kellogg of the RAND Corporation, Leroy R. Alldredge of the Operations Research Office of the Johns Hopkins University, and Michael Ference, Jr., of the Ford Motor Company alone needed briefing. Van Allen began by summarizing the outcome of a conference he and Porter had attended in late February at NRL to discuss plans with Hagen, Rosen, and Newell. The conferees had agreed that the objectives of the program were, first, to place an object in orbit and prove by observation that it was there; second, to obtain a precision optical track for geodetic and high-altitude atmospheric drag measurements; and, third, to perform experiments with internal instrumentation. After achieving the second objective in one or two flights, the third goal would take precedence over the second. Rosen had emphasized the necessity of keeping satellite weight to a minimum until such time as the vehicle proved able to carry more. "If necessary to buy improved performance by reduction of payload," the best way, the five men thought, would probably be to start flights with an empty third-stage bottle 18 inches in diameter by 50 inches in length, try next a 6-pound payload consisting solely of the Minitrack instrumentation in a minimum size capsule, next an 8.5-pound payload in a 20-inch sphere, and thereafter payloads ranging from 14 to 18.5 pounds with 2 to 5.5 pounds allowed for scientific instrumentation. Larger loads might be feasible later. The Laboratory would supply experimenters with "black-box" specifications for payload capsules, and a list of pertinent Minitrack characteristics. The salient features of the telemetry system were not yet determined.[11]

These terms left open the possibility of using a cylindrical package of instruments, if not the cylindrical or conical outer body which Van Allen and Hagen had wanted, but the chances looked slight of getting as much as eight pounds of instrumentation into any IGY satellite. Van Allen and George Ludwig, in an attempt to meet the Vanguard specifications, had already scaled down their plan for cosmic ray observations, but the severe restrictions on weight automatically knocked out several otherwise useful proposals submitted at the Ann Arbor symposium.[12]

With these conditions in mind the working group turned to establishing the criteria that should govern the selection of onboard experiments. First was that of scientific importance, "to be measured by the extent to which the proposed observations, if successful, would contribute to the clarification and understanding of large bodies of phenomena . . . and/or be likely to lead to the discovery of new phenomena"; the second was that of technical feasibility as established by use of similar techniques in rockets or other scientific vehicles, by the "adaptability of the instrumentation to the physical conditions and data transmission potentialities of presently planned satellites," by "the nature of data to be expected," and by "feasibility of interpretation of observations into fundamental data"; the third was that of the competence of the persons or agencies making proposals, an assessment based on past achievements in work of the kind proposed; and the fourth was that of the necessity, or strong desirability, of using as the vehicle for the experiment a satellite rather than a sounding rocket or a balloon. The group, however, went on record as wanting to encourage proposals that would help develop a "reservoir" of scientific competence in devising experiments for future satellite flights "even though such work may not yield practical apparatus for the short-range IGY program." [13] The WGII, in short, saw its task as extending to plans for space exploration long after the IGY ended.

With long-term objectives in view, the working group also decided "to give further consideration to the establishment of a worldwide net of telemetering receiving stations for the continuous or nearly continuous reception of observed data" and "to consider concerted action on development of solar batteries, telemetering systems of more general applicability, data storage and read-out devices." The main business of the session, however, was the preparation of a preliminary listing of priorities among the experiments already submitted, despite the virtual certainty that the months ahead would bring in a number of new propositions, some of which might fill existing gaps in the fields of inquiry thus far covered. The consensus ran that the selection of internal instrumentation was lagging behind other parts of the satellite program; since the apparatus for every experiment chosen would have to undergo rigorous laboratory tests and, if possible, flight tests in rockets in order to check its capacity to withstand the vibration, the shock of accelerating velocity, and the environmental conditions to be encountered in the vacuum of space, the least complicated instruments might well take at least two years to perfect. There was no time to lose. So the WGII then and there ranked the proposals before it, putting in order of choice the experiments that appeared to have most scientific value and be best suited to early satellite flights, and, in a second, "B," category those that might be flown later but were as yet of doubtful utility or feasibility.[14]

After appraising the WGII's report, the panel voted to limit for the time being "the positive standing on the Priority Listing" to nine projects, thereby discarding three. The subsequent voluntary withdrawal of two others further reduced the number. By common agreement, Friedman's Lyman-alpha experiment, the environmental studies, and the proposal from the State University of Iowa headed the list from the beginning, even though sounding rocket experts at NRL had pointed out that a series of rocket probes could provide cosmic-ray observations as well as could an instrumented satellite.[15] The plan of Van Allen and Ludwig by now called for apparatus that was to consist of two parts: instruments for continuous transmission of signals marking the instantaneous intensity of cosmic rays registered by the Geiger-counter, and, second, equipment to store the instantaneous intensity data during each orbit for read-out on command over the Minitrack stations. A small cylinder would house batteries, a receiver, a transmitter, tuning forks, a tape recorder driven by a ratchet system, scalers, and generators. The Geiger-Mueller tube would project about 4.5 inches from the top plate of the cylinder.

Fourth on the priority list was an experiment entitled "Measurement of Interplanetary Matter," submitted by Maurice Dubin, E. R. Manring, and others of the Geophysics Research Directorate at the Air Force Cambridge Research Center. Their plan was to detect the spatial distribution and size of particles colliding with the satellite—even those as small as one micron in diameter—by recording the acoustical energy generated on impact. Instrumentation would consist of a sensitive piezoelectric transducer on the inside surface of the satellite shell, a transistorized amplifier, a storage device, a power package, and a time-delay switch set to operate after the Minitrack telemetry began to transmit. The memory device was to count the number of stored impacts, record the distribution of particles by size, and transmit the information when the amplifier was in use. Somewhat similar in purpose to Singer's rather simple meteoric dust erosion experiment, Dubin's appeared to have greater scientific utility; perhaps the two might be combined. The panel recommended that Dubin receive the grant of $89,045 that he requested, Singer, though his plan stood in category B, a grant of $47,150.

At the top of the B list was a proposal for meteorological observations, prepared by William G. Stroud of the Signal Corps Engineering Laboratories (SCEL). Its primary objective was to measure the global distribution and movement of cloud cover and to relate it to the gross meteorology of the earth. Contrasts in terms of sunlight reflected from cloud, sea, and land masses, as viewed from a spinning satellite during the telemetry time, should furnish the basic data. Two photocells using a single telemetry channel would look out in diametrically opposite directions at a known angle to the spin axis of the satellite. The signals from the photosensitive cells would be stored in an airborne magnetic tape recorder and, at interrogation, be played back during a one-minute interval over a one-watt transmitter. A switch would turn off the equipment during periods of darkness and turn it on again when the satellite reemerged into sunlight. Although the question would later arise as to whether the data could be transmitted in form that lent itself to meaningful scientific interpretation, the panel recommended a grant of $93,000 to develop the apparatus.[16]

An experiment of great scientific interest but of somewhat doubtful technical feasibility was H. E. Hinteregger's proposal to develop photoelectric techniques for study of extreme ultraviolet solar radiation. A member of the Geophysics Research Directorate of the Air Force Cambridge Research Center, Hinteregger hoped to trace the high yields of photoelectric emission in this range of radiation. Although the probability of adapting the equipment to an IGY satellite looked small, the WGII and the panel believed the plan worth encouraging. An award of the $5,000 Hinteregger requested would keep the total figure recommended for grants to date to $371,320. The panel therefore could allot $275,000 for NRL experiments and keep $570,000 for pending projects without exhausting the $1,262,000 earmarked in the satellite budget for internal instrumentation.[17]

During spring 1956 several men tendered proposals for ionospheric studies, but all of them called for use of ground station receivers and airborne transmitters with radio frequencies incompatible with the Minitrack's 108 megacycles. While suggesting that the authors discuss possible compromises with John Mengel, the WGII undertook to remind all experimenters that their equipment must not interfere with Minitrack and should rely on the Vanguard telemetry system. Telemetry time during an initial orbit would be only thirty seconds, although after the orbit was determined, the time could be increased to two or three minutes by changing the positioning of the ground station antennas. A Vanguard development still in the tentative stage might provide a sixteen-channel memory circuit for storage of data during orbit and a read-out system responsive to command from telemetry at the ground stations. Magnetic tape running at a constant speed would record the signals at ground stations and playback tape would give the data on wide film or paper strips with time markers.[18] Experimenters who received grants, the panel decreed, must be informed promptly of their chances of having their apparatus flown during the IGY. The working group had decided to assign each high priority project to a particular vehicle; if a bird failed to orbit, the next vehicle launched would carry the experiment, but, if a second failure occurred, the untried experiment would have to yield to the next on the list. A prototype of the apparatus for every project in the A category should be ready by January 1957 for WGII approval.[19]

Of the three propositions the WGII added before the end of 1956 to those tentatively chosen earlier, one was an experiment called "Geomagnetic Measurements" prepared by James Heppner and colleagues at NRL. In essence it was the magnetometer experiment outlined in NRL's original presentation to the Stewart Committee, later described in a paper at the Ann Arbor symposium, and now reworked to accommodate it to the small payload of a Vanguard satellite. Its objectives were, first, to gauge the intensity of the earth's main magnetic field during magnetic storms and measure its contribution to the total storm disturbance as a function of time and latitude; and, secondarily, "to determine the existence or nonexistence of extraterrestrial currents during the initial phase of a magnetic storm and to improve our knowledge of ionospheric currents giving rise to diurnal and irregular variations of the magnetic fields, especially near the magnetic equator." The principal instrument was to be a nuclear magnetic-resonance magnetometer in the form of a coil around a sample of liquid that contained a high proportion of protons. A magnetic field would be produced by passing a current through the coil, thus polarizing the protons' magnetic moment. Upon cutoff of the polarizing current, the proton moments would precess about the earth's field at a frequency determined by the field's strength and induce a voltage at that frequency in the coil. This signal, following amplification, would be fed to the telemetering transmitter for transmission on command to the Minitrack stations. At each Minitrack station there was to be "a proton precessional magnetometer to simultaneously measure accurately the total scalar field, declination and inclination." The paucity of observation time and the lack of a recording and storage device in the satellite led the working group to question the value of the attainable results, but Van Allen, after displaying a model of a magnetometer built and used successfully in rocket flights by the State University of Iowa, testified to the probable workability of the special model under design by Varian Associates of California. With this experiment approved for funding, the sum allowed for NRL's three onboard projects rose to $597,000.[20]

The second new experiment the WGII recommended—rather hesitantly, to be sure—was a plan largely worked out by William O'Sullivan of the staff of the National Advisory Committee for Aeronautics (NACA) at Langley Field, Virginia. It required no scientific instruments and little equipment other than a gas-filled bottle and a device to eject a thirty-inch inflatable sphere from the satellite at the moment of third-stage burnout. It was designed to permit optical observers to compute air densities and to measure atmospheric drag on the aluminum-foil-covered plastic body. If the perigee of the satellite were less than 200 miles, the life of the sphere would be extremely short, but as the weight of sphere, gas tank, valve, and ejection trigger together would not exceed nine ounces, the paraphernalia could go into a satellite carrying fairly heavy instrumentation for another experiment. NACA would meet most of the costs.[21]

Verner E. Suomi of the University of Wisconsin proposed the third experiment recommended at the end of 1956. The objective of Suomi's experiment, known as "Radiation Balance of the Earth," was to measure the long-wave radiation emitted from the earth, from direct sunlight, and from sunlight reflected from the earth, and also the short-wave radiation reflected from the earth and either shielded from or insensitive to the other radiations. Harry Wexler of the Weather Bureau, who had submitted a somewhat similar but more complicated plan, warmly supported Suomi's proposal as scientifically important and technically feasible. It should supply means of charting the gains and losses in the earth's heat budget during a satellite's lifetime. Of four small thermistor sensors mounted on the ends of the satellite antennas, one sensor would be sensitive only to long-wave radiation emitted by the earth, the second equally sensitive to other types of radiation, and the third and fourth sensors only to short-wave radiation reflected from the earth. During the satellite's orbit a selector switch would monitor each sensor and feed signals from a coding oscillator to each for a preselected time. An airborne magnetic tape recorder would store the data until, upon passage of the bird over a Minitrack ground station, a command turn-on signal initiated the playback sequence. While the intricacy of the instrumentation militated against the chances of its being ready for use during the IGY, the panel recommended an initial grant of $50,000.[22]

Although priorities necessarily would change if tests of instrumentation so dictated, in February 1957 the panel made its selection of experiments to fly in the first full-size Vanguard satellites. Assuming four successful shots during the IGY, the WGII had proposed to assign a "package" containing two experiments to each of the first three birds, a single experiment to the fourth. The panel concurred. Package I was to take the equipment for the environmental studies and the Lyman-alpha experiment. Package II was to contain the apparatus for Van Allen's cosmic ray observations and either for Dubin's measurements of interplanetary matter or for Singer's, of meteoric dust erosion, provided either of those could employ a masked photocell instead of the radioactive method. By April progress on Dubin's instrumentation captured for it the coveted place in package II. Package III was to carry the instruments for Heppner's geomagnetic measurements and O'Sullivan's inflatable sphere. For package IV the panel wavered between Stroud's cloud-cover experiment and Suomi's radiation balance. The upshot was a decision to let both proceed until the work was further advanced and then request the country's leading meteorologists to name the more useful of the two.[23]

Although the expense of developing internal instrumentation was beginning to run unexpectedly high, the panel assigned "back-up" status to several projects. So Hinteregger's scheme of measuring extreme ultraviolet solar radiation won official endorsement and, later flown in a rocket, produced some significant data. Singer got funds to complete his radioactive meteor erosion gauge, money which the panel switched in 1958 to support his endeavor to devise means of determining the electrostatic charge accumulated by a satellite, but he never submitted detailed designs or an experimental prototype. A grant went also to the group of men, headed by William Pickering, at the Jet Propulsion Laboratory for development of instrumentation to measure the integrated light from various parts of the celestial sphere, using a set of color filters and a photomultiplier detector. Planned for use in case more experiments could be flown during the IGY than anticipated, the JPL equipment was never put to the test in a satellite, but the work on it proved useful in preparing later projects. An experiment proposed by Martin Pomerantz of the Bartol Research Foundation and Gerhardt Groetzinger of the Research Institute for Advanced Studies was in turn given funds, even though the ion chamber and circuitry designed to identify the heavy primary cosmic ray nucleii and the possible variations in their flux appeared unlikely to be available for IGY satellites.

Interestingly enough, no experiment in the life sciences received endorsement. Yet in 1951 and 1952 Kaplan had viewed the possibilities of medical research in the aeropause as an impelling reason for a satellite program, and the NRL proposal to the Stewart Committee had alluded to the feasibility and utility of studying the behavior of living cells in the vacuum of space. Early in 1957, a biologist at the National Institutes of Health submitted a plan for recording the effects on yeast cells placed in an orbiting satellite, but the panel postponed action on the idea.

All told, the panel rejected seventeen proposals and, counting those dealing with tracking and engineering problems, approved over twenty before October 1957.[24]

In backing experiments too heavy or too elaborate for Vanguard satellites or adapted primarily to space probes in rockets, the TPESP was adhering to the principle announced by the IGY National Committee at the end of 1956. If, as the committee and the panel assumed, the scientific exploration of space continued after the IGY was over, technological advances would surely supply the bigger vehicles needed for the purpose. Indeed as early as May 1957 the panel had reason to think the time near when larger satellites than the twenty-inch Vanguard sphere could be circling the earth, for Pickering reported that the Army already had available a launcher capable of putting a thirty-pound payload—which would include the rocket casing —into an orbit of 1,000-mile apogee and 200-mile perigee. Pickering knew whereof he spoke, inasmuch as the Jet Propulsion Laboratory under an Army contract had been working closely with the Army Ballistic Missile Agency on developing the Jupiter-C rocket. Despite Vanguard's configuration and the relatively slow spin rate of its last stage, at least one package of satellite instrumentation, notably that for the Van Allen experiment, could be fairly easily adapted to flight in the Army vehicle. When Porter asked why the Defense Department did not sanction use of the new rocket as a backup for the NRL–Martin launcher, Paul Smith explained that the DoD had considered the plan but vetoed it as needless; Vanguard tests were on schedule and satisfactory.[25]

Work on the satellite structures and instrumentation meanwhile had moved along rapidly at NRL. Since every experimenter was to furnish his own apparatus, the naïve reader might assume that the team at the Laboratory would have relatively little to do: merely supply the satellite shell, the telemetry, the antennas, and the tracking transmitter, and then install the package of experiments. But those tasks in themselves were formidable. The satellite as planned had to carry a device for separating the sphere from the rocket casing after third-stage burnout; the shell and every mechanism in it must be sturdy and lightweight; and to fit all the items into a twenty-inch sphere required miniaturization of an order never before thought attainable. The layout of instrumentation, furthermore, had to vary from one satellite to the next so as to adapt each to the particular experiments it was to accommodate. Nor did the job end there. Thermal control presented enormous difficulties, and the entire testing program demanded scientific knowledge, great ingenuity, and endless patience.

Common background simplified the dealings of the Vanguard scientific unit with the authors of the IGY experiments, for, like most of the latter, a number of men at the Laboratory, as pioneers in space exploration with sounding rockets, had had to design and build their own instruments in the past. Mutual respect and cordial relations between the two groups could not, however, greatly lessen the steadily mounting burden of work carried on for Vanguard by NRL scientists. Homer Newell and his deputy, John Townsend, who were in overall charge and directed the program through a so-named Satellite Steering Committee, accordingly asked Robert W. Stroup early in 1957 to serve as general coordinator and trouble-shooter. In late April 1957 the steering committee, with Hagen's approval, arranged a three-day conference which brought all the experimenters to the Laboratory where they could see the work in progress, observe the testing arrangements, and discuss their individual needs and quandaries.[26]

There Robert C. Baumann, head of the mechanics and structural unit, described the separation device under manufacture by the Raymond Engineering Laboratory of Middletown, Connecticut. He also displayed models of the twenty-inch satellite shell that Brooks and Perkins of Detroit were spinning from flat sheets of magnesium into two hemispheres which skilled craftsmen at NRL then riveted together; a small trap door gave access to the interior. The mechanical features of the folding antennas to be affixed to the outer surface were largely of Baumann's design; the electronic parts were the work of Martin J. Votaw and Roger Easton; NRL shop hands were building the arrays. Mechanics in the shop were also fabricating the 6.4-inch "grapefruit" to be flown in the test vehicles. For the satellites carrying packages I, II, and IV, the original scheme of an inner sphere to house the scientific instrumentation had given way to a cylindrical container attached to the outer shell by a spider framework of tubular metal; teflon-covered supports shielded the pot from the grids. The satellite carrying the magnetometer experiments, on the other hand, was to have a different configuration: a thirteen-inch fiberglass sphere with a fiberglass stem projecting from it several inches to support the sensor. At a later date the steering committee would discover the necessity of providing a spin-reduction mechanism for the satellites flying the cloud-cover experiment and the NACA inflatable sphere in order to allow, in one case, a longer scanning interval for the photocells and, in the second case, ample time to inflate the plastic subsatellite.[27]

Whitney Matthews, who was in charge of the electronic layout within the spheres, demonstrated the general scheme of stacking the layers of "cards" of miniature mechanisms and locating their power supply. As was true of other features of the satellites, each package would differ somewhat in both content and arrangement from every other. Although little of the work was in final form in April 1957, the economy of the intricate layouts, the complexity of the tiny parts, and the delicacy of the workmanship were already plainly visible. Roger Easton then explained in detail how the tracking and telemetry equipment would function. The tracking signals would be amplitude modulated for telemetering the scientific data obtained from the satellite-borne detectors. A transistorized transmitter in the satellite weighed 1.25 pounds, including the weight of Minitrack batteries for two weeks' operation, and used about 7.5 pounds of mercury cell batteries that would give three weeks of continuous operation at fifty milliwatts output for telemetry. Telemetering might be continuous or could function on command. When commanded, a receiver weighing twelve ounces including its power supply would pick up the signals sent from the ground. While the command receiver in every satellite would be standard, the telemetry transmitter might differ in type, depending on the requirements of the experiments carried. On the ground the telemetry receiver was to be located at some distance from the tracking receiver. NRL had built the first tracking transmitters in its shop; testing and evaluation of performance had been going on at the Blossom Point station since July 1956.

Of equal or perhaps even greater interest to the visiting scientists attending the conference were the accounts of the methods under development to provide thermal protection for the satellite shell and its payload. Solving these problems, above all those deriving from the effects of radiation under various conditions, called for pooling the talents and experience of several men, notably Hermann LaGow, who had planned the environmental studies accepted for the first satellite flight, Richard Tousey and Louis Drummeter of the NRL Optics Division, Milton Schach of the Electronics Division, and George Hass of the Engineer Research and Development Laboratories at Fort Belvoir.

Tousey had made some of the first calculations in the fall of 1955, contributing his knowledge of optics to ensure that protective coatings on the exterior of the booster and on the satellite shell would have sufficient reflectivity to permit telescopic observation of the course of the rocket as it rose and then optical acquisition and tracking of the satellite in space.[28] Schach undertook the "thermal design," that is, the calculations of what temperatures to expect at various points in the satellite's orbit, in darkness and in daylight, the selection of the optimum thickness of coating materials to emphasize their emissiveness of solar heat radiation, and methods of keeping the satellite's surface free of contaminating substances such as soot which would ultimately raise the satellite's temperature. Hass worked out the techniques of applying the successive surface coatings—the gold plating, the chromium evaporated to vapor and deposited to serve as a primer, the silicon oxide to serve as a barrier, the thin layer of evaporated aluminum to give a mirror-like finish, and finally a film of silicon oxide to control emitted radiation. Drummeter and Schach were chiefly responsible for developing the sunlight simulator with carbon arcs as the source of high-intensity light. Through windows in the large cylindrical vacuum tank in which the coated sphere sat for two or three days of testing, the simulated sunlight beat upon the satellite's surface and indirectly heated the inner pot of instruments. Measurements of the effects furnished means of determining the most desirable material and thickness of the layering required. LaGow acted as advisor and monitor on all these operations. Every man concerned with temperature control worked closely with every other.[29]

Newell and Townsend, who had initially objected to launching a 6.4-inch satellite with no internal instrumentation, were reconciled to the plan as thermal testing proceeded, for the agreement to place temperature sensors on the 6.4-inch shell promised to verify the findings of the Laboratory tests or else supply data that would permit development of better thermal control for the larger instrumented satellites. Although the research and testing was still going on when the conference with IGY experimenters took place in the spring of 1957, the NRL thermal experts were already fairly confident that they could limit temperature changes within the instrument container to some two or three degrees during any one orbit. While expectations ran that a Vanguard satellite would have only a few weeks' life, the possibility of its lasting longer led to endeavors to adapt thermal controls to seasonal as well as diurnal changes.

The men attending the conference received, moreover, thorough indoctrination in the standards of performance which the Vanguard group demanded of every experimenter's equipment. If, when put through the whole gamut of tests at NRL, his instruments could not withstand extremely high random and sinusoidal vibrations, changes of temperature ranging from 0°

127

J. Paul Walsh (left), Deputy Project Director of Vanguard, and Homer Newell participate in a conference in the director's office.

to 60° C, and a simulation of the sudden acceleration that would occur when the second-stage rocket separated from the first or the second from the third, then the Satellite Steering Committee could reject the experiment outright, unless the originator was able to correct the weaknesses thus revealed. He was to test his work carefully in his own laboratory before sending his instrumentation on to NRL. If possible it should also be tried out in a rocket flight. After testing each item, the Vanguard group would need a minimum of three months to check the reliability of the assembled package. The precious space in a satellite must not be wasted on faulty scientific paraphernalia. All electronic and experimental equipment should have a life of at least 1,000 hours.[30]

At Van Allen's request, a special session explored the progress of the development of dependable solar batteries, especially the work going on at the Signal Corps Engineering Laboratory. Using chemical batteries, Vanguard satellites and most of the onboard apparatus thus far proposed would have an active life of only a few weeks. Satellites with longer life consequently needed a solar battery system for long-term power. All experiments would have far greater value if they could operate for several months. Although the SCEL during the past year had solved a number of problems, others remained, notably the sharp decline of open circuit voltage when temperatures rose from 20° to 80° C. A new type of cell, however, might provide an answer. Use of clusters of solar batteries, moreover, might charge a low-voltage secondary battery, while transistor d.c.–d.c. converters could supply the higher voltages, such as the twenty-three volts needed for the Minitrack transmitter. Every experimenter undertook to supply the SCEL with a statement of his power requirements. On the whole, the prospects looked bright for having usable solar batteries available before the end of the IGY. But if so, and if the batteries extended observing time significantly, "then the radio tracking, telemetering, data analysis and computation items in the budget must be correspondingly increased." To prepare for that contingency, the Academy's satellite panel believed $200,000 necessary in the immediate future.[31]

John Hagen had earlier suggested that every experimenter or one of his associates should be present at NRL during "the final preparation period" of his apparatus. In any event, two months in advance of a flight he must send the Laboratory an instruction manual explaining how his instruments would work; the field crews would need the manual to learn how to set up the recording mechanism. During the conference, each team of experimenters met separately with the Vanguard staff to draft an explicit agreement about what services the Laboratory would perform, what the outside scientists were to be responsible for. Each team, moreover, reported on the then status of its project. Van Allen and Ludwig expected to have the entire package of their instruments in the Laboratory's test rooms by August. The tape recorder, a source of trouble earlier, was now functioning smoothly, the circuitry working well. Photographs showed the 9-inch cylinder 4.5 inches in diameter containing eight modules encapsulated in foam to provide mechanical rigidity and thermal insulation, the electrical system, and the overall layout. The package weighed 13.41 pounds, the framework 7.09. Testing of the instrumentation for the Lyman-alpha experiment and the environmental studies was also well advanced. Progress on other projects was somewhat slower.[32]

When the impending start of the IGY brought a number of internationally known scientists to the National Academy in June 1957, the presence of several eminent Russian astronomers and geophysicists added greatly to the interest of the occasion. Contrary to later popular hearsay in the United States, the Soviets talked of their plans, and I. P. Bardin turned over to Lloyd Berkner a document entitled "U.S.S.R. Rocket and Earth-Satellite Program for the IGY." In the section of the exhibit hall given over to the satellite program, reporters clustered around John Hagen and his Russian counterpart. Hagen in answering questions repeatedly spoke of the NRL satellite, whereupon a very junior member of the IGY staff corrected him with "The National Academy's satellite, Dr. Hagen."[33] The incident revealed the constant stress the Academy felt obliged to put on the nonmilitary character of the program. July publication of the first issue of the *IGY Bulletin* served again to remind readers that the National Academy was responsible for the undertaking.

The summer of 1957 was not a time of rejoicing for the men handling satellite finances. The expenses of the Glenn L. Martin Company and sub-contractors had increased steadily since October 1956, as indeed had the costs of the scientific parts of the program. Despite the transfer of $5.5 million of National Science Foundation funds to NRL in October 1956 and another $1.862 million in March 1957, the Vanguard comptroller estimated in April that the bill for the entire satellite program would run to $110 million, NRL's costs alone to $96.162 million.[34] The Navy budget was not the direct concern of the IGY satellite panel, but it would become so if financial exigencies caused serious slippages in the Vanguard launching schedules. As every setback to the program dimmed the chances of the Academy's winning endorsement of its cherished plans for twelve shots, the USNC secretariat awaited with anxiety the results of the Navy's appeal to Congress.

Thanks to inaugurating in September 1956 a new financial reporting system which required the Martin Company and other NRL contractors to submit detailed cost data monthly, Thomas Jenkins, the Vanguard comptroller, was able to refine earlier estimates; the Laboratory was going to need $34.2 million more than was then available to see the satellite job through to completion. Rather than ask for piecemeal allotments, the Defense Department and the Bureau of the Budget concluded that the wiser course was to seek authorization from Congress to turn over to the Navy the whole amount in a lump sum.[35] Jenkins tabulated the figures for the congressional committees:

FINANCIAL SUMMARY [36] (In thousands of dollars)

	Total Funds Required	Funding of and Agency Responsible for Program Costs		
		NRL	Other DoD	NSF (NAS)
A. Available				
1. DoD emergency fund	46,300	46,300	x	x
2. Miscellaneous	2,500	2,500	x	x
3. DoD direct funding-logistic support	4,411	x	4,411	x
4. Estimated range-use charge by Air Force	4,227	x	4,227	x
5. NSF 1956 Supplemental	18,362	13,126	x	5,236
Total	75,800	61,926	8,638	5,236
B. 1958 request to Congress	34,200	34,200	x	x
Total	110,000	96,126	8,638	5,236

His figures were all-inclusive, a fact rarely understood, then or later, by people not intimately involved with Vanguard. From the cost of the new radar, the blockhouse, and telemetry equipment—all destined to serve Cape Canaveral for years—to the pay of NRL shop hands for part-time work on Vanguard hardware, every iota of expense was taken into account, even items that a less meticulous person might think properly chargeable to Laboratory or Navy overhead.

The accompanying text gave no precise explanations of why costs for the vehicle, estimated at $28.1 million in March 1956, had risen in fourteen months to $57.111 million, or why the Navy's overall costs, including its work on radio tracking, telemetry, data reduction, and the satellite itself, now in May 1957 seemed certain to exceed $96 million. Yet at hearings in August the Senate Committee was on the whole astonishingly amenable, in spite of nearly universal confusion among committee members over the differences between sums voted to the Science Foundation for the IGY and funds allocated to the Navy for the same program. "We appropriate money to the National Science Foundation," said Senator Magnuson, "and then we appropriate extra money to them for the International Geophysical Year, of which they then gave you some . . . . Now the Navy is asking for extra money for their part of the Vanguard program, which is part of the International Geophysical Year." John Hagen simply replied: "It never has been very straight in the record." Ten days later Congress authorized the Secretary of Defense to release to the Navy the $34.2 million requested. The hand-to-mouth financing of the previous two years need no longer hamper Project Vanguard.[37]

But money alone could not solve the Laboratory's problems. A measure of the discouragement pervading NRL as summer turned into autumn was an exchange between Rosen and Hagen. "John," said the technical director despairingly, "we're never going to make it in time," to which the older man replied gently: "Never mind! It's a good program, worth following through."[38] At Cape Canaveral the TV-2, originally scheduled for flight tests in June, had not left the launch pad at the beginning of October.[39]

While Richard Porter and the IGY staff at the Academy were aware of the successive delays, when CSAGI gathered in Washington on 30 September for a week-long conference on rockets and satellites, most members of the TPESP knew relatively little about Vanguard tribulations. The panel had not met since 1 May. At that time news emanating from the Pentagon had been blandly reassuring. Now panel members learned that the flight test of a Vanguard test vehicle with two dummy stages and minus a satellite was set for mid-October. The panel meeting held on 3 October was thinly attended: Lyman Spitzer had resigned; Odishaw, Spilhaus, and Newell were engaged with the CSAGI sessions; Van Allen was en route to the South Pacific.

Chairman Porter was worried, but if the other men present shared his un-spoken belief that a Russian satellite was nearly ready for launching, they kept their foreboding to themselves. Most of the discussion focused on optical tracking and how to speed up deliveries of the Baker-Nunn cameras. Whipple, to be sure, raised the question of whether the Academy was satisfied with the Vanguard flight schedules, but Porter pointed out that launchings were solely a DoD responsibility. The panel adjourned without pursuing the subject.[40] Twenty-four hours later everyone even remotely interested in the American program was asking when the United States would put its first satellite into orbit.

8
CREATING A HOME ON THE RANGE

DURING the closing months of 1955 and well into the new year the process of siting Project Vanguard checkout and launch facilities at the Air Force Missile Test Center (AFMTC) at Cape Canaveral proved nearly as troublesome, discouraging, and time-consuming as the negotiating of the prime contract between NRL and Martin.

In the fall of 1955 the 15,000-acre missile firing range on the snake-infested and palmetto-covered sand dunes of the Florida flatlands was completing its sixth year as the Long Range Proving Ground for American guided-missile development. Congress had established the range for this purpose in 1949. In 1950 the Department of Defense had assigned responsibility for its operation to the United States Air Force, and had named it the Air Force Missile Test Center (AFMTC). By the end of the following year the Air Force had set up administrative and telemetry headquarters eighteen miles south of the range at a former coast guard and seaplane base just south of the village of Cocoa Beach, to be known henceforth as Mason M. Patrick Air Force Base (PAFB). Three guided missiles had lofted from the Cape, and the range had received the official designation AFMTC that it would retain throughout the lifetime of Project Vanguard. The press called it "Cape Canaveral" and Air Force and Vanguard men often spoke of it as the Atlantic Missile Range (AMR),[1] the official designation it later received.

Vanguard's request to use the DoD launching site, aired unofficially in September 1955 and cast in formal form a few weeks later,[2] elicited no cheers from AFMTC management or its parent body, the Air Research and Development Command (ARDC). On 2 December, ARDC Headquarters approved of Vanguard's request "in principle," [3] but made it abundantly clear that the Air Force viewed with alarm the prospect of making room at a high-priority military installation for a no-priority scientific program. Vanguard's eventual acceptance at the range depended upon its ability to work out with AFMTC a modus operandi consistent with the National Security Council's order that the earth satellite project be so conducted that "it does not materially delay other major defense programs." [4] Nor did ARDC's statement in December [5] that Vanguard could communicate directly with the field do more than smooth somewhat a rough path, since on an informal basis project representatives had been communicating vociferously with the men in charge at Cape Canaveral for many weeks.

Since early fall, conferences at Patrick Air Force Base or conferences elsewhere relative to field problems had been taking place at frequent intervals. During September Vanguard's telemetry boss, Mazur, made several exploratory trips to the Cape, accompanied by Captain C. B. Ausfahl, an Air Force officer attached to the Naval Laboratory, and Alton E. ("Al") Jones, one of NRL's bright young men, whose calculations prior to the submission to the Stewart Committee of the NRL satellite proposal had been instrumental in the decision of the Laboratory and GLM to use two liquid-fuel rockets rather than one in the Vanguard vehicle. [6]

At Patrick the NRL team talked at length with high-ranking officers and with a number of engineers working for Pan American Airways, industrial contractor charged with servicing AFMTC operations, or for Radio Corporation of America, subcontractor responsible for field instrumentation. It was information gathered during these preliminary discussions that later impelled the Laboratory, in the face of objections from both the Florida test center and the Martin Company, to install at the Cape the newly developed radar antennas and data-acquisition equipment that would enable Vanguard to fly PPM/AM telemetry packages in the first stage of its vehicles. [7]

Although the NRL investigators' primary purpose in Florida was to survey the instrumentation available at the Cape, they also made inquiries as to the form in which the Vanguard people should prepare and present to AFMTC a list of the facilities and ground support equipment they would need at the range to put up their birds. One of their first discoveries was that where material to be provided by the Air Force was concerned, the test center preferred a statement as to the accuracies desired rather than the names of particular items. For a time at least this idiosyncracy on the part of the missile management was a stumbling block to old NRL hands, accustomed to asking their procurement unit for what they wanted by name or by general description without as a rule bothering to explain what it was supposed to accomplish.

"One of our problems in the early Vanguard days," Mazur would confess later with characteristic forthrightness—

was that we simply didn't know how to deal with a paper-type organization like the Air Force. Those of us at NRL had got our rocket experience flying Vikings at White Sands. That was a relatively informal and leisurely program. When we needed this item or that we got on the phone or sent a note up to the Lab people in Washington, telling them to buy the damn thing or to whack it together themselves and send it down to us. We didn't have to write up a thousand documents in quintuplicate, as we soon found we had to do for the Air Force. What we asked for in a hurry we got in a hurry because, as we were fond of telling the fly boys in those days, at NRL procurement existed for the Laboratory, not the Laboratory for procurement. The Air Force procurement cycle was a good deal slower, and for us at any rate a tough nut to crack. Quite frankly we might have ended up lacking many of the things we vitally needed at the field had it not been for the good offices of Colonel Gibbs.[8]

Mazur's reference is to Lieutenant Colonel Asa B. Gibbs, who in fall 1955 was director of tests at AFMTC. It was a happy day for the scientific satellite effort when in late September Gibbs was relieved of his duties as test officer and ordered to "report aboard" the Vanguard management staff as Air Force liaison or project officer.[9] Although the heavy-set, cigar-puffing colonel impressed members of the satellite group as jovial,[10] he could be effectively tough when it came to procedures he considered important. Gibbs knew his way around the Air Force. Throughout what Mazur called the "agonizing period," [11] the drawnout struggle to fit Project Vanguard into the Cape Canaveral picture, he made life considerably easier for the planning staffs by educating them in the ins and outs of the command setup at Patrick. It was a complicated one, and the planners escaped many irritating delays because of Gibbs' skill in opening channels of communication between Vanguard control center at NRL in Washington and Base headquarters at Cocoa, Florida.[12] Gibbs also served the project as a gadfly. Repeatedly he scolded at NRL and Martin, urging them not only to put their list of requirements in a form acceptable to AFMTC, but to make certain it covered every conceivable item, "right down to the number of rolls of toilet paper you may eventually need." The conscientious project officer neglected to specify how the Vanguard people were to delineate the "accuracies" of this homely necessity; he did warn them, solemnly, that unless their list of requirements was complete they might come right up to the point of firing their first test vehicle, only to find that many of the items of equipment and supply essential to the undertaking were simply not on hand.[13]

Drafting the requirements list was a Martin responsibility, but in actual practice it became a joint effort since NRL had to supply the designers and engineers at the big aircraft plant in Middle River with a vast amount of information. Hundreds of man-hours of work—and talk—went into the preparation of the document. In its finished form it reflected every predictable aspect of the satellite operation, ranging from what those involved hoped to learn from the flight of each vehicle to the probable behavior aloft of all vehicles and scientific payloads. A preliminary list submitted to the range command in November 1955 brought an immediate demand for extensive revisions. To make sure that these were properly executed, Colonel Gibbs organized a team of NRL and Martin experts and arranged for them to spend a week at Patrick Air Force Base, working directly with range officials and technicians. The outcome of this effort was a respectable document labeled "Test Program for Vanguard Launching Operations." Satisfactory though it was, even this "final" list was tentative. Immediately after its acceptance by the range command in May, Project Director Hagen issued a directive outlining the procedures for making anticipated "changes, deletions," and "additions." [14]

The work of preparing the requirements list proceeded concurrently with other developments. ARDC's approval "in principle" of Vanguard's bid to use the Cape gave the project a foot in the door. The actual opening of the door required additional and in some cases ticklish maneuvers. These began to come to a head in mid-December during a four-day conference at PAFB attended by representatives of the Base, the Martin Company, and the Laboratory. Heading the Air Force contingent was slight, dapper, outspoken Major General Donald N. Yates, USAF, commander at AFMTC. A newcomer to the Vanguard forces was Commander, later Captain, Winfred E. Berg, who only a few hours before had joined Vanguard as official Navy representative and senior project officer. Slender and striking under a plume of thick black hair that the problems of Vanguard would quickly sprinkle with silver, Berg was no stranger to the oldtimers at NRL. He had worked with them at the Laboratory, leaving in the early 1950s to take a staff position in the Comptroller's Office at the Pentagon as technical adviser on missiles and related fields to Assistant Secretary of Defense Wilfred J. McNeil. As Vanguard's chief trouble-shooter from 1956 on—its "fire brigade" to use his own expression—Berg would enjoy ample opportunity to deploy his notable talents for cutting through red tape and getting things done.[15]

During the opening phase of the mid-December conference at Patrick, General Yates and his aides confirmed a worrisome fact, already spelled out in Air Force correspondence. To test and launch its hardware Vanguard must have a variety of facilities, notably a launching pad and adjoining blockhouse, and a hangar where it could assemble its three-stage vehicles and maintain offices, laboratories, and storage bins. For the time being all such installations on the Cape, including some under construction, or about to be, were assigned to high-priority military programs. To use the Cape, Vanguard must either construct its own launch complex and hangar or arrange to share facilities with some existing program.

The meeting warmed up when, on top of this revelation, the NRL-

Martin representatives informed the Air Force Commander that they were planning to launch their first test vehicle during the coming June. Yates' blunt reaction was "Ridiculous! It'll take you that long to get the money." At this point Commander Berg took over for Vanguard. Would the Air Force support use of the range, he inquired, if by the first of January, only two weeks hence, assurances were in General Yates' hands that the requisite funds would be available as needed? Yates knew a sporting proposition when he heard one. "It's a deal," he snapped, and Berg hastened back to Washington to make swift use of his connections at his old stamping grounds, the Comptroller's Office in the Defense Department. His negotiations were well-advanced before he realized that it might be impossible to make good his end of the deal with Yates because the first of January was a holiday, with all government offices closed. He went ahead, hoping that either the desired assurances would reach him prior to the first or that Yates would overlook an insignificant delay. On the second he was able to inform the AFMTC commander that the authorization of funds was processing satisfactorily. Yates accepted this as fulfillment of their agreement. A week later formal authorization came from the Department of Defense, and Vanguard was assured a home on the missile range.[16]

The chronic afflictions of the scientific satellite effort, too little time and too little money, accounted for the decision not to erect a new launch complex but to opt for space in one already up or soon to go up. In 1956 the rocket team headed by von Braun for ABMA, the Army Ballistic Missile Agency, was in the midst of a five-year flight program at the Cape, testing versions of the Army's Redstone rocket, including the Jupiter-A and Jupiter-C intermediate-range ballistic missiles. Acting for Vanguard, Berg, Rosen, Mazur, and Roger Easton visited both Cape Canaveral and the ABMA offices in Alabama in an effort to obtain quarters at the Redstone launch complex on a joint-use basis. The facts uncovered during these sessions were discouraging. The Redstone firing schedule at AFMTC was too tight to make a sharing arrangement feasible. In other words the Redstone people said no, and the Vanguard negotiators had to look elsewhere. Fortunately the Air Force had recently undertaken the development of the Thor, another intermediate-range missile, with Douglas Aircraft Company as prime contractor responsible for design and fabrication. No firing installations for the Thor were in existence as yet at AFMTC, but plans were underway for the construction of two blockhouses and four pads on what would shortly be labeled launch complexes 17 and 18 on the southern rim of the easternmost bulge of the Cape. The Thor program did not call for a flight from complex 18 for several years—a situation that enabled General Yates to accede to Milton Rosen's request in January 1956 that Vanguard occupy some of the facilities projected for that complex, sharing the blockhouse with the Thor project and testing and firing its vehicles from an adjoining launch area, subsequently designated as 18A.[17] It was understood that the use of these facilities and any changes required in them were to be financed with Project Vanguard funds.

Shortage of time and money also entered into the decision not to build a new gantry crane for servicing the Vanguard vehicle but to adapt to Vanguard use the one the Laboratory rocket crew had recently developed but never used for its now abandoned Viking program at White Sands Proving Grounds in New Mexico. At a request from the Navy, the Jacksonville (Florida) District of the Corps of Engineers agreed to supervise the transfer of the ninety-five-foot Viking gantry to its new home. In April the Corps negotiated a contract under which the Treadwell Construction Company, of Midland, Pennsylvania, designers and makers of the Viking gantry, undertook to dismantle the crane at White Sands and see that it was reassembled and ready for use at AFMTC during the forthcoming September. The contract directed Treadwell to provide the crane with an additional working platform and other modifications, and the Air Force took steps to install at complex 18A a 225-foot railroad track on which the huge service tower could be moved up to and away from the vehicle-launching structure.[18]

The crane was not the only Viking hardware to become part of the scientific satellite development. During March a group of GLM and Laboratory men spent six days at White Sands examining the ground support equipment left over from the Viking program and tagging for shipment eastward those items that they believed Vanguard could utilize.[19] Efforts to make do with what was "on the shelf" and other penny-pinching stratagems were a commonplace of the program. The usually stately, although at times opaque, official correspondence yields a plea from the Martin Company "that we be allowed to use a spring removed from a Victor Mouse Trap, manufactured by the Animal Trap Company, Lititz, Pa., for our magnetic disconnect doors in the second and third stages of the Project Vanguard vehicles. . . . Allowing us to use this spring will save considerable time . . . . The spring will be cadmium plated with an Iridite Number 1 finish . . . ."[20] Signed by D. J. Markarian, Vanguard Project engineer for GLM, this letter was dated 18 April 1956. A decade later John T. Mengel, NRL's radio-tracking genius, could still recall with a chuckle a trip to Sears Roebuck and Company to purchase 200 inexpensive screw-jacks for use on his multi-million-dollar antenna arrays.[21]

To AFMTC as the Defense Department's agent fell the task of providing the permanent field facilities Vanguard would require. In February the Secretary of Defense authorized the Center to have built for the project a combination office and vehicle-assembly building, to be designated as Hangar S and located four and a half miles northwest of launch complex 18.

Blockhouse used by Project Vanguard at Cape Canaveral. The horizon window overlooks the launch pad.

Word from the Corps of Engineers, responsible for the supervision of the construction work, was that the hangar would not be ready until April 1957, at the earliest. Meanwhile, Vanguard was to occupy half of an existing building, Hangar C, in the vicinity of the old lighthouse some two miles northeast of the Thor launch areas.[22] Plans approved by the base put all the Thor pads 600 feet from the blockhouse. This was the standard distance, acceptable to range safety officers bent on making sure that the men assigned to the remote control room in the blockhouse during firings enjoyed as much protection as possible. Some groans arose at the range safety office

139

when Vanguard requested a separate pad only 200 feet from the blockhouse. Following a few conferences, however, the Base command in the person of General Yates bowed to Vanguard's argument that the shorter distance would be a worthwhile economy since it would eliminate the "repeater" equipment necessary for the longer line-runs of a 600-foot installation. When in April the Corps of Engineers employed the J. A. Jones Construction Company of Charlotte, North Carolina, to build the six-million-dollar Thor launch complexes, an amendment to the contract required Jones to provide a simplified pad for Vanguard at a point about midway between the Thor-Vanguard blockhouse and Thor pad 18A.[23]

The made-to-order pad was one of several concessions Vanguard succeeded in obtaining from the Air Force. NRL and Martin insisted also on a "wet pad," one equipped with a plumbing system capable of supplying water for cooling the flame duct of the launch structure and for other purposes connected with static and flight firing tests. For Yates this request posed problems. It meant piping off the main line water intended for the Thor project. It also called for the emplacement of a spilloff basin for catching the water poured through the flame duct, and Yates feared that a basin in the Vanguard launch area would create later difficulties for the Air Force's IRBM program. He agreed to the wet pad only after receiving assurances from NRL and Martin that no interference with the Thor schedule would ensue.[24]

During the many conferences at Patrick, some of the Vanguard spokesmen reacted with what one of them called "incomprehension" to the punctilious manner in which General Yates carried out his duties as commanding officer of the range. In a report covering a lively session in February, Robert L. Schlechter, GLM's field manager for Vanguard, regaled his colleagues at Middle River with a picture of the General "fulminating" into the meeting room to assert that "some structural monstrosity Douglas [the Thor manufacturer] proposed" building on Complex 18 "would be erected over his corpus delicti." According to Schlechter, the general then asked Commander Berg what the Vanguard people would do if their desire for special facilities on the Thor complex were not met. Schlechter quoted Berg as replying that if "this were the case," Vanguard would make another effort to share the Redstone facility. "After further questioning," the eloquent Martin engineer's account continued, "General Yates learned that we still proposed to static fire on the Redstone pad in the event of sharing." This would have required a major modification of the Redstone complex, a prospect that brought violent repercussions. Schlechter's recollection was that "the well known matter hit the fan, with General Yates bringing his corpus delicti into the picture again. Whereupon the U.S. Navy told him he would have to accept this situation if directed by the Department of

Defense. There was some profanity from the general, but the Navy held its ground." [25]

But if tempers flared occasionally in the heat of battle, the ultimate consensus in Vanguard circles was that on the whole General Yates was a good friend to the scientific satellite program. It was Yates who slipped into the contract covering erection of the Thor facilities a clause ordering the Jones Construction Company to complete the blockhouse on complex 18, the one scheduled for Vanguard use, two weeks ahead of the one on the other Thor complex. Nobody noticed this effort to speed things up for the satellite-orbiting program until November 1956, when glass for the heavy "horizon-seeing" windows arrived at the Cape. In the words of a Project official, "all hell broke loose" when the builder discovered that for the time being he had only enough glass for one window, and proceeded to put it into the Vanguard blockhouse in accordance with contract.

"Yates didn't always take kindly to our propositions," Berg has commented,

> but he was open minded. He was responsible for providing service to all of the projects at AFMTC. He had no choice but to honor their priorities, but he did a wonderful job of finding ways of helping his one no-priority program, Vanguard, to meet its commitments. In the beginning, for example, he was most reluctant to let us install at AFMTC the then new high-precision radar, AN/FPS-16, for the purpose of tracking the Vanguard vehicle during and immediately after the launch sequence. He contended that the Azusa tracking system already there was quite adequate, and that he didn't want every new project 'clobbering up my range' with its own special equipment. Later, when he saw how effective our tracking system was, he made some changes in his own and went out of his way to point out that ours had proved an asset. AFMTC still uses the Azusa system as improved by Yates, but it also uses the AN/FPS-16 brought in by Vanguard.[26]

Late spring 1956 found site-clearing crews gathering at Cape Canaveral to begin work on the Vanguard launch complex and hangar. NRL and Martin had established a series of "beneficial occupancy" dates, a schedule best described as a demonstration of the unending triumph of human hope over the facts of life. First the national steel strike of that summer, then a spate of local strikes created what scientists and engineers quaintly insist on calling "slippages," meaning delays.[27]

Since AFMTC's responsibility for construction extended only to permanent facilities, the Vanguard managers asked Martin to provide the "Static and Flight Firing Structure" the Vanguard field hands would need to test-fire and launch their vehicles. Beginning in January, Martin designers rapidly produced a set of acceptable drawings. The structure they called for consisted essentially of a flame deflector tube surmounted by a steel platform $9\frac{1}{2}$ feet high and capable of supporting two removable rocket stands. One of these was for use during static or free firings of either the first stage of the Vanguard vehicle or of the entire vehicle, the other during static firing tests of the second stage only. In all, the structure consisted of five major components: deflector tube, tube support, test platform, plumbing system, and rocket stands. There were also access stairs and such facilities as a shower and a fireman's pole-type escape hatch for the protection of men working on the erected rocket.[28]

In July both NRL and Martin experienced an uneasy weekend when an examination of blueprints revealed a discrepancy between the water outlet the Jones Construction Company had installed in the Vanguard pad and the corresponding water inlet on the launching structure, then still in the New York plant of Loewy Hydropress, the subcontractor charged with its manufacture. The opening of the manhole-like outlet in the pad was thirty-six inches, that of the inlet on the structure thirty inches, and the interphase—the elbow created by the makers of the structure to link the two units—was thirty inches at both ends. The situation was potentially troublesome. Vanguard construction had already suffered serious slippages. In an effort to hold these to a minimum the Jones gang at Cape Canaveral was working at top speed. Discovery of the discrepancy came on a Friday, and Jones was planning to pour concrete on the following Monday for that portion of the pad wherein the water outlet was located.

Commander Berg came to the rescue. At his urgent request, Jones agreed to delay the concrete pouring for one day. Berg then put in a phone call to John Manning, deputy engineering services officer for production at the Naval Laboratory in Washington. On Saturday and Sunday Manning's shop crew cast a 1,700-pound, thirty- by thirty-six-inch reducer to replace the unusable interphase. Early Monday morning the new elbow was on a plane. That evening it was at Cape Canaveral, ready to be installed. "A miracle!" said Berg. "Nothing of the sort," retorted shop-boss Manning. "At NRL doing the impossible is routine procedure." [29]

Since Project Vanguard was a step into the unknown, frequent changes of policy were inevitable. In the beginning the project managers thought of their launching program as consisting of two major phases, a test phase and a mission phase. For the test phase they had asked the Martin Company to manufacture at least seven vehicles—TV–1 through TV–5, plus two or more backup vehicles. For the mission phase, they called for at least six satellite-launching vehicles, SLV–1 through SLV–6.[30] Their plan, in those early days, was to use the first four test vehicles for vehicle-testing purposes only. Not until the time came to launch TV–4 did they intend to try sending a satellite into orbit. Soon after the project started, however, those in charge began changing these plans. By the summer of 1957, the Naval

Research Laboratory had informed the Martin Company that from TV–3 on, all Vanguard vehicles were to have satellite-bearing capacities, and the company had reoriented its production plans accordingly. Meanwhile, the scientists and engineers working in the so-called "ballroom" at NRL had developed the "grapefruit" satellite, having a diameter of only 6.4 inches and weighing only 4 pounds, much smaller than the 20-inch, 21.5-pound sphere the project managers had previously contemplated for use in their first attempt at orbit.

The official record fails to reveal the reasons behind these changes, but the memories of the men responsible for them supply the deficiency. The complete success of the launch of TV–1 in May 1957 had demonstrated the capacity of the new solid-propellant third-stage rocket, flown then for the first time, to meet all of its requirements. In addition the excellent performance of the previously untried spin stabilization system removed the necessity of carrying a heavily instrumented load on TV–3. In July 1957, therefore, the project planners decided to substitute the smaller six-inch spherical package. Outfitted with a beacon transmitter, it was to perform two functions: one, test the downrange instrumentation that the Vanguard people had installed to convert the Atlantic Missile Range into a satellite launching range and, two, lighten the load on TV–3 so that if all went well in this first test of the entire three-stage vehicle, a satellite would result. Current or foreseeable slippages in the launch schedule prompted the decision to begin trying to orbit a satellite with TV–3 instead of waiting until TV–4. The project managers reasoned that unless they attempted to fly a satellite earlier than planned they might find themselves unable to make good on their commitment to put one in orbit during the forthcoming International Geophysical Year. From Donald J. Markarian, who served as Martin's project engineer for Vanguard, comes a convincing summary of the thinking behind the development of the grapefruit satellite: "We had to put more instrumentation in the test vehicles than in the mission or satellite-launching vehicles. We were confident that the lighter SLV—the mission vehicle—would launch a 21.5-pound ball into orbit, but we were doubtful about the ability of the heavier TV—the test vehicle—to do so. Common sense suggested that we'd be better off succeeding with a smaller satellite than failing with a bigger one." [31]

Recurrent difficulties notwithstanding, Vanguard entered the fall of 1956 in fair condition. To be sure its hopes of launching a rocket during the preceding summer had gone glimmering. There was reason to believe, however, that its adjusted schedule calling for an initial firing in December was realizable. By the end of October an advance unit of the field crew had set up shop in the project's temporary hangar near the old lighthouse, the launching complex was far enough along to be used, the access roads to it were being completed, the gantry crane, freshened by a coat of green paint, stood ready for use, and the first of Vanguard's test and launch vehicles, TV–0, had arrived at Hangar C and was undergoing checkout.[32] Progress was relatively satisfactory also in other aspects of the program, notably in those connected with tracking and with the acquisition of tracking and telemetered data.

9

THE TRACKING SYSTEMS

Tracking, as the term indicates, means measuring the position of a moving object, natural or man-made. Optical tracking with sighting instruments is as old as astronomy, which is at least as old as written history. A child of modern science, radio tracking using radar, radio direction finders, Dovap, and other electronic schemes, emerged during the first quarter of the twentieth century. Radio interferometry, the technique employed in Vanguard's electronic tracking system, entered the picture in the 1940s. Like most of the electronic tracking techniques then in use, this one required the presence of a signal source, a transmitter, in the object being tracked. Employing two receiving points on the ground and comparing the phases of the signals each of them separately received from the airborne source, radio interferometry had the advantage of yielding highly accurate angles.

It achieved practical form in 1948 when engineers with Consolidated Vultee Aircraft Corporation (Convair) created for the Army the Azusa tracking system, using an interferometer.[1] Simultaneously with this development, NRL scientists were working with underwater sound interferometers. The two groups of experimenters were in close contact. They frequently exchanged ideas, and in the early 1950s Milton Rosen and his NRL crew at White Sands built and field tested a tracking system, using the radio interferometer principle, for application to ballistic missile guidance for the Viking rocket.[2]

When in early 1955 Hagen's people, specifically Rosen and his colleagues, began drawing up their plans for launching an earth satellite, they devoted much thought to the problem of tracking so small an object. Most upper-air research scientists advised them to rely on optical tracking. It was the tried and true method. More to the point, Fred L. Whipple's camera observations of meteorites entering the earth's atmosphere had demonstrated in the late 1940s that modern terrestrial optical instruments could "spot" an object weighing only a few kilograms and moving at a substantial distance.[3]

Rosen had doubts. At his request Richard Tousey of NRL checked Whipple's visibility computations. He confirmed these calculations, but found in them no answer to an important question: Granted optical instruments could see the satellite if they could find it in the first place—but could they find it? In Tousey's opinion their chances of doing so were only one in a million.[4] There was the further consideration that optical tracking, although highly accurate, has limitations. The best of sighting instruments can pick up a satellite only when the sun is five degrees below the horizon—that is, at dusk and dawn—and even then only under certain weather conditions.[5] Convinced from the beginning that NRL must look elsewhere for an adequate satellite-acquisition method, Rosen had asked John T. Mengel and his NRL Tracking and Guidance Branch to develop an electronic system for use in conjunction with an optical one.

For guidance purposes, the Azusa system had performed satisfactorily at White Sands and elsewhere. For tracking a satellite, however, it was out of the question since it required an airborne transmitter far too large for a small scientific payload weighing no more than thirty pounds, if that much. Refinements and modifications were indicated, and Mengel and his assistants shortly came up with an arrangement that, although based on the Viking radio interferometry techniques, required instead of a heavy transponder only a thirteen-ounce transmitter [6] and employed different operating frequencies and antenna configurations. It was in essence a new system.

Mengel's name for the system, Minitrack, derived from the system's utilization within the satellite of "an oscillator of minimum size and weight . . . to illuminate pairs of antennas at a ground station which measures the angular positions of the satellites using phase-comparison techniques." [7] As eventually developed, the thirteen-ounce Minitrack oscillator, quartz-crystal controlled and fully transistorized, had a ten-milliwatt output, operated on a fixed frequency of 108 megacycles, and had a predicted lifetime of ten to fourteen days.[8]

In a series of papers and speeches prepared over a two-year period beginning in 1956, Mengel and his colleagues—notably Roger L. Easton, his assistant at NRL, and Paul Herget, director of the Cincinnati Observatory and a consultant to Project Vanguard—assessed the role of tracking in the satellite program and described some of the characteristics of the Minitrack. In March of 1956 Mengel told a group of scientists and engineers:

The final realization of man's efforts to place a satellite in orbit about the earth will immediately pose a new series of problems: how to deter-

146

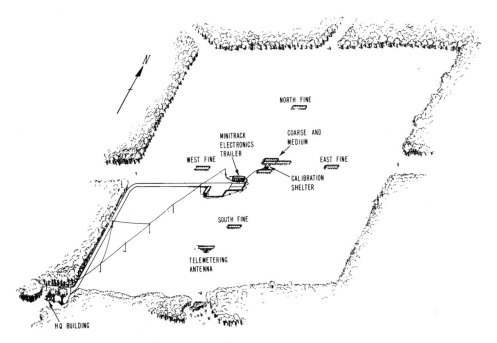

Schematic of a typical Minitrack station.

mine the precise orbit that it is following; and how to measure what is happening within the satellite from the vantage point of a ground station. The immensity of the first of these programs, how to prove that the satellite is in fact orbiting—the acquisition phase—can be realized by . . . an analogy . . . . Let a jet plane pass overhead at 60,000 feet at the speed of sound, let the pilot eject a golf ball, and now let the plane vanish. The apparent size and speed of this golf ball will closely approximate the size and speed of a satellite 3 feet in diameter, at a height of 300 miles . . . . The acquisition problem is to locate the object under these conditions, and the tracking problem is to measure its angular position and angular rate with sufficient accuracy to alert non-acquiring tracking stations, those trying to follow the satellite by optical means, as to the time and position of expected passage of the object.[9]

In a popular article,[10] Mengel and Herget likened the antennas of Vanguard's Minitrack stations to human ears. "An individual," they pointed out, "locates the source of sound by virtue of the phase differences in the sound waves, arriving at different times at his two ears. Similarly the listening units of the minitrack system are pairs of receiving antennas, set a measured distance apart, which indicate the direction of the signal by phase differences in the radio waves . . . ." They described the reception pattern of a Minitrack station as a fan-shaped beam making an arc of "100 degrees north and south and 10 degrees east and west. In the long direction of the beam we have three pairs of antennas, spaced respectively 500, 64, and 12 feet apart; in the narrower east-west dimension, two pairs with spacing of 500 and 64 feet. The north-south and east-west [antenna arrays] give us two angles which [combine to indicate] the actual direction of the satellite."

The Minitrack stations, as they came into being, showed minor variations. Generally speaking, however, their major components were the fixed arrays for angle tracking, one fixed antenna array for telemetering reception, a rhombic communications antenna, a ground station electronics trailer, a telemetering trailer, a communications trailer, and associate power sources and maintenance units. These components required about twenty-three acres with a minimum gradient of the land less than one degree in the region of the angle tracking antenna arrays. Those in charge of choosing the sites took care to place each station where the adjacent terrain did not exceed an elevation angle of ten degrees for at least half a mile, twenty degrees for five miles. They also took care to select a site at least two miles from heavy electric power installations and at least five miles from airports or airways.[11]

The inclusion of Minitrack in the satellite-launching proposal NRL submitted to the Stewart Committee had had direct bearing on the committee's decision in August 1955, to accept the Laboratory's proposal in preference to the Army-Orbiter proposal, which originally contained no provision for electronic tracking.[12] The NRL proposal also mentioned in passing the creation at the Laboratory of a less elaborate version of the Minitrack system. Developed by Roger Easton, this abbreviated system would come to be known officially as the "Mark II," unofficially as the "Jiffy" or "Poor Man's" Minitrack. Using radio-frequency phase-comparison techniques by means of hybrid junctions, the Mark II later became the nucleus of "Project Moonbeam," a program sponsored by Project Vanguard to encourage radio hams and their organizations to build their own stations and participate in tracking satellites.

There were two forms of the Mark II, known, respectively, as the "simple" and the "advanced." The amateurs who joined Project Moonbeam used only the simple form. Costing about $5,000 to erect as against twice that much for the advanced Mark II, this arrangement consisted of two matched antennas in an extended base array, a receiver, and oscillograph. Passage of the artificial earth satellite produced a pattern of reinforcement and cancellation, successively, of the signals received at each antenna. These were recorded as a pattern of peaks and nulls. A difficulty in interpreting these records, even where satisfactory time signals were recorded in an auxiliary channel, was in determining which null corresponded to the time of the passage of the satellite through the principal plane of the antenna array. In general these ambiguities were resolved at the Vanguard satellite computing centers by references to the data received at the prime Minitrack stations operated by professionals.[13]

As soon as Project Vanguard became official in the fall of 1955, the individuals charged with tracking began work on its implementation. In view of the technical hurdles to be cleared and the problems incident to coordinating the work of numerous military units, university laboratories, individual experts, private industries, and elements of the world scientific community, the progress of the electronic tracking group was remarkably smooth. Years later Mengel would be able to recall "a few personality clashes," but in his opinion these were "par for the course for a program that made use of some of the best astronomers in the country." In addition, the middle months of 1957 saw two slippages: the late arrival of construction material at one of the Minitrack installations and a one-month delay in completing the communications network set up to tie together all of the far-flung elements of the system.[14]

In April 1956, the technical panel's Working Group on Tracking and Computation (WGTC), stamped "approved" on the plans for the optical tracking system, as drawn up by the Smithsonian Astrophysical Observatory, that is, by Whipple and J. Allen Hynek. Simultaneously the group gave its blessing to Whipple's budgetary estimates, an action that soon thereafter had the effect of setting aside $3,380,610 for the optical tracking program.[15] The plans prepared for the optical system envisaged the use of two separate but cooperating groups, one to acquire or "find" the satellite, the other to track it. Acquisition was the responsibility of the group known as Project Moonwatch. Whipple's initial announcement of his intention to use amateurs in this fashion brought expressions of skepticism from members of the technical panel. "Some of my colleagues," the feisty, personable SAO director would recall later, "were convinced that too few amateurs would volunteer, and that those who did would not always perform satisfactorily. Time, I'm happy to say, has proved these fears to be groundless. The amateurs joined up in droves all around the globe. They did a splendid job for Project Vanguard, and they have been doing an increasingly more effective one for the American space effort ever since."

The nucleus of the other phase of the optical system, the precision tracking phase, would consist of the twelve observation stations, set up around the world and operated by professionals. Each station was to have a high-precision camera and associated clock. It was assumed that reduced data obtained from these installations would permit the calculation of definitive orbits for use in correlating with satellite-borne and ground-based experiments, thus providing valuable scientific information. Early in the planning period staff members of the Smithsonian Astrophysical Observatory undertook to develop a list of possible camera locations. On

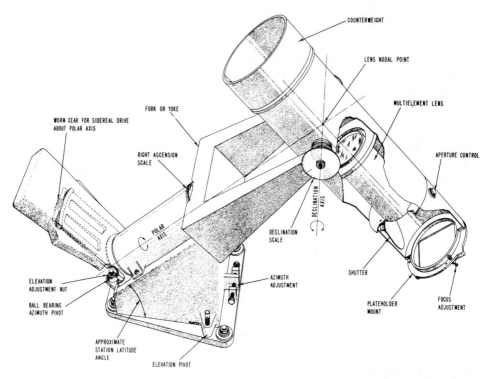

COUNTERWEIGHT

LENS NODAL POINT

MULTIELEMENT LENS

FORK OR YOKE

WORM GEAR FOR SIDEREAL DRIVE
ABOUT POLAR AXIS

RIGHT ASCENSION
SCALE

APERTURE CONTROL

DECLINATION
AXIS

POLAR
AXIS

DECLINATION
SCALE

ELEVATION
ADJUSTMENT NUT

AZIMUTH
ADJUSTMENT

SHUTTER

BALL BEARING
AZIMUTH PIVOT

PLATEHOLDER
MOUNT

FOCUS
ADJUSTMENT

APPROXIMATE
STATION LATITUDE
ANGLE

ELEVATION PIVOT

The Baker-Nunn satellite tracking camera and diagram of its components.

visits to more than a score of countries, Hynek and other SAO scientists met with local scientists and government representatives to work out collaborative methods for setting up and operating the units. In the beginning thought was given to placing all of them at Minitrack stations. This arrangement would have saved money, but it turned out to be impractical because of the differing requirements of the two systems. "A camera station," Whipple has explained, "needs clear skies, whereas the principal need at a radio tracking station is a flat surface away from noise. The Minitrack station in Ecuador, for example, was ideally located for radio tracking purposes, but it stood in an area where the skies are overcast most of the time. Joining our camera stations with the Minitrack network would have spared us many logistic headaches, but generally speaking we just couldn't do it." In the end an optical and a Minitrack station were combined at only one point, Woomera, Australia. Of the remaining optical locations, two were in the continental United States—at Jupiter, Florida, and Organ Pass, New Mexico. The others were at Olifansfontein, Union of South Africa; Cadiz, Spain; Mitaka, Japan; Naini Tal, India; Arequipa,

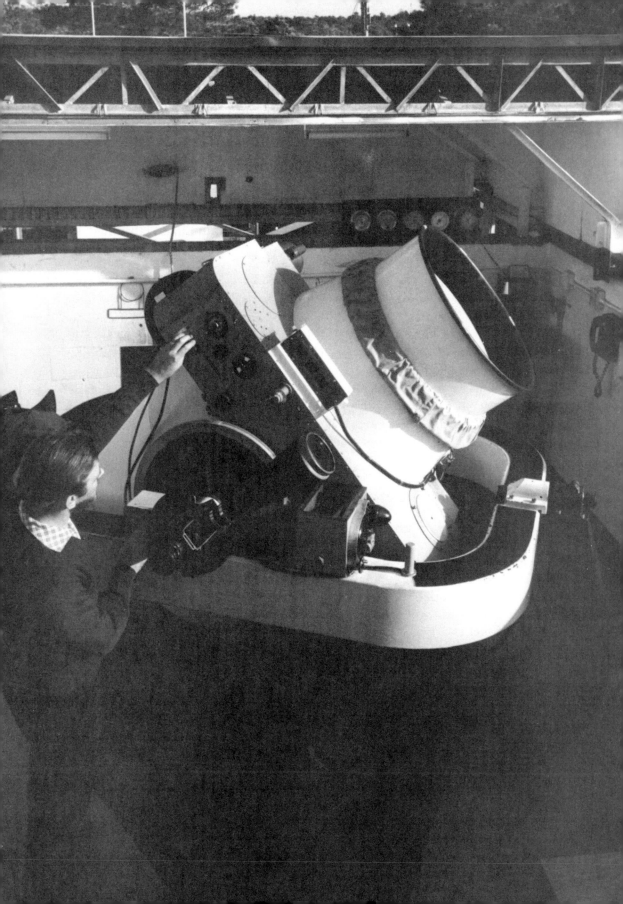

Peru; Shiraz, Iran; Curaçao, Netherlands West Indies; Villa Dolores, Argentina; and Haleakala, Maui, Hawaii.[16]

To cope with the unprecedented task of picking up a tiny man-made satellite orbiting at a great altitude, Whipple and Hynek supervised the development of the Baker-Nunn high-precision telescopic camera with an unusually large aperture.

The Baker-Nunn took its name from its principal creators: James G. Baker, consultant to the Perkin-Elmer Corporation of Norwalk, Connecticut, and Joseph Nunn of South Pasadena, California. Baker designed the camera, Nunn, its mechanical elements. Perkin-Elmer fabricated the optics, and the Boller and Chivens Company of South Pasadena built the camera proper.[17]

Although the performance of the cameras was destined to fulfill expectations,[18] their production was plagued by setbacks. The October 1957 meeting of the technical panel at IGY headquarters in Washington found the scientists present discussing the lagging camera-delivery schedule with such heat that for once lean-faced, high-domed Fred Whipple lost his calm. Why so much fuss, he wondered, over the delays in the optical tracking program? Was the panel satisfied with the Vanguard launching schedule? His reference, of course, was to the inability of the NRL-Martin field crew to meet its flight-firing dates.

Panel chairman Porter took care of Whipple's query with his statement that the launching schedule was a Defense Department responsibility, not a panel responsibility. With a sharpness that the formal phraseology of the minutes fails to disguise, he added gratuitously that for Whipple's information the radio tracking program, also a Defense Department responsibility, was practically on schedule.

Whipple promptly changed the subject, or rather shifted it to different grounds. He described himself as miffed by a recent newspaper story charging that delays in the camera-delivery schedule were holding up the entire Vanguard project. Porter agreed that the newspaper story was in error. He offered to so inform the panel's parent body, the IGY committee, but the panel as a whole took no action on this suggestion and the matter was dropped.

Whipple reported that the first Baker-Nunn camera had been assembled and was under test at Pasadena. It would go to its station in a few weeks. A second camera was expected to arrive at Pasadena for assembling and tests by early December. Whipple anticipated that after that the "optics would be completed to allow for camera completion at about one-month intervals." In time he hoped to cut this to three weeks. Clock production was good. The same could be said for the station-building schedule. Basic materials had been shipped to six of the twelve stations, although recently the ship bearing material to Japan had caught fire—damage as yet undetermined.

Porter lost no time in getting to the heart of Whipple's summary. It was plain that the precision-camera network would not be in full operation until August 1958. At that point only four months would remain of the International Geophysical Year as then projected—the period during which Project Vanguard was committed to the launching of at least one satellite.

Porter concluded his gloomy observations by raising—and in effect answering—a pointed question. Should the camera program simply be canceled? In his opinion the answer depended on whether the satellite program did or did not continue beyond the termination of the IGY year. As to that, no one could yet say. A general discussion followed, at the end of which the sense of the panel found its way into the minutes. It was that the members viewed the slipping "delivery schedule of the cameras with grave concern." They had even considered their elimination, but had concluded that such a step would be inadvisable. Implicit in this action, as in so many of the actions of the panel scientists, was an abiding faith that Project Vanguard was only the beginning of a long American space program. Their final word on the camera-delivery problem was that the IGY committee should "adopt every means in its power to expedite the schedule." [19]

In this respect, regrettably, the powers of the National Committee were limited by the severe technical problems involved in the production and testing of a highly advanced optical camera. Observations at the first completed station, Organ Pass, would not begin until November 1957. Not until June 1958—only a month ahead of the date so dolefully forecast by Porter—would the entire precision-camera network be in operation.[20]

The delay would have been even greater had not Porter visited the plant of the fabricators of the camera optics, Perkin-Elmer, where he learned that the Perkin-Elmer people had underestimated the job, bid too low, and were reluctant about lavishing expensive overtime on an undertaking that was going to leave them with a loss. Having discovered the trouble, Porter persuaded the Smithsonian Institution to renegotiate its contract, raising the price to a point where Perkin-Elmer could break even. This was an intricate transaction, calling as it did for coordination with the National Science Foundation, the Smithsonian Astrophysical Observatory, and the American IGY Committee; but it was accomplished rapidly and had the effect of accelerating the camera-production program considerably.

Lest the delays in establishing the optical network take on more significance than they should, it is worthwhile anticipating a little to point out that its long-run contributions to the scientific satellite project were substantial. Nor do the Technical Panel minutes reflect all of the difficulties confronting the supervisors of the optical program. Actual release of the funds allotted to them was a function of the Smithsonian Institution, and in the 1950s the administrative procedures of the Smithsonian were not geared to a fast-moving project like Vanguard. As a result, Whipple and his associates in Cambridge had to spend an undue amount of time cajoling the firms with which they dealt into supplying them with material and services on credit while they waited for the creaking fiscal wheels in Washington to revolve. Project director Hagen sagely characterized the optical program as a "prudent" backup to NRL's radio tracking system.[21] The program also engendered great interest in the project as a whole by its sponsorship of Project Moonwatch. This visual observing program— "visual" rather than "optical" since its members made no use of cameras— gave amateur astronomers around the world the opportunity of playing a useful role in the Vanguard tracking system. Organized and directed by the Smithsonian Astrophysical Observatory, Project Moonwatch received as much if not more attention from the press than any other aspect of Project Vanguard. Newspaper readers were intrigued by the picture of hundreds of small groups of enthusiastic star-gazers getting together in open fields or on lonely hilltops in an effort to "catch the satellite" with their binoculars and small telescopes.

Guided by instructions from SAO, the amateurs gathered preliminary orbital data on the satellites. These they transmitted to Cambridge, whence they were distributed in the form of ephemerides to the technicians orbital data on the satellites. These they transmitted to Cambridge, whence

As Whipple had anticipated, the announcement of the formation of Moonwatch in early 1956 brought an enormous response. Visual observation teams sprang up in North America, South America, Africa, Europe, and Asia, in the Middle East and at such remote specks on the map as Station C and Fletcher's Ice Island T–3 in the Arctic Basin. Before the Vanguard program terminated, 250 teams with approximately 8,000 members were functioning. Teams were organized by universities, high schools, government agencies, commercial organizations, private science clubs, and groups of laymen. The United States alone accounted for 126 groups.

The volunteers furnished their own equipment. As the program took form, however, SAO succeeded in obtaining some special equipment from army surplus. The observatory sent these items to some of the more effective groups.[22] For a time, ironically, the success of Project Moonwatch was a source of embarrassment to the Smithsonian Astrophysical Observatory. The minutes of the technical panel find Whipple complaining that the unanticipated growth of the program was putting a severe strain on his small administrative staff.[23] So much for the penalties of glamor.

In projecting the electronic tracking program, during the early Vanguard days, the Naval Research Laboratory contemplated using only four Minitrack stations along with a prototype station, but by the time Vanguard was ready to issue its first full-scale report of progress in December 1955, the radio tracking experts were thinking in far more elaborate terms.[24] In its final form the Minitrack network would consist of fourteen ground installations.

One of them, the prototype Minitrack station at Blossom Point Proving Ground in Maryland, forty miles south of Washington, was a service-station for the other elements of the network. Here the electronics engineers, the foreign scientists, and other technicians chosen to operate the stations received the bulk of their training. Blossom Point also provided a center for the development of system tests and of procedures for calibrating the Minitrack antennas, principally by using ground-based cameras to photograph aircraft carrying Minitrack test transmitters and ground-controlled flashing lights against a background of stars.[25]

Three stations, set up immediately downrange from Cape Canaveral on the islands of Grand Bahama, Antigua, and Grand Turk, functioned in connection with a fourth unit, a radar installation at Patrick Air Force Base, to keep tabs on the Vanguard vehicle during launch and shortly after the third-stage powered-flight phase of the launch sequence. Supplemented by a station in South Africa and another in Australia, the remaining elements, the so-called prime Minitrack stations, were strung out in a north-south line along the east coast of North America and the west coast of South America so as to form a "picket line" across the expected path of all satellites launched from Cape Canaveral. Mengel was confident that with this arrangement—this "fence of" stations located generally along the seventy-fifth meridian in the northern and southern hemisphere [26]—"we have a 90 percent chance of intercepting every pass of a satellite which is higher than 300 miles." [27] In accordance with a decision reached in December 1955, all prime Minitrack stations and some of the subsidiary units of the network included high-gain (TLM 18) antennas for gathering telemetered data from the satellite.[28]

All branches of the military contributed to the erection, operation, and maintenance of the radio tracking system. Three Army agencies—the Corps of Engineers and its Army Map Service and the Signal Corps—took responsibility for most of the construction work and for setting up the communications network. The Bureau of Yards and Docks, a Navy unit, obtained use of the necessary lands for the Prototype Station at Blossom Point, and the Air Force arranged for the installation at PAFB and Grand Bahama Island of the two high-precision tracking radars, the XN–1 and XN–2 models of the AN/FPS–16.[29]

With assistance from native scientists, Army men operated the five

Interior of the Minitrack tracking van at Blossom Point, Maryland.

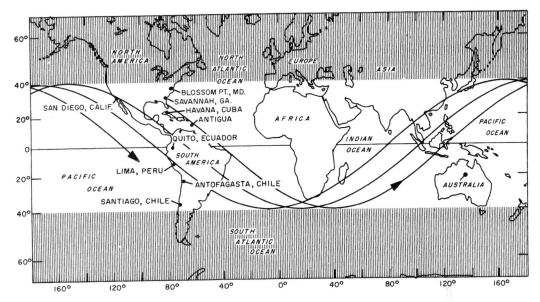

The network of primary Minitrack stations, as of 25 January 1957.

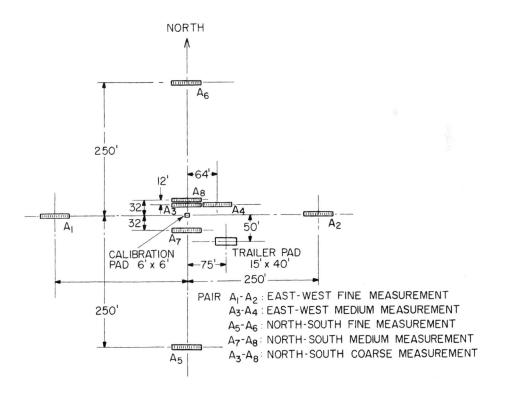

PAIR A_1-A_2 : EAST-WEST FINE MEASUREMENT
A_3-A_4 : EAST-WEST MEDIUM MEASUREMENT
A_5-A_6 : NORTH-SOUTH FINE MEASUREMENT
A_7-A_8 : NORTH-SOUTH MEDIUM MEASUREMENT
A_3-A_8 : NORTH-SOUTH COARSE MEASUREMENT

Schematic of a typical Minitrack antenna system.

prime Minitrack stations in Latin America. They also operated the prime Minitrack station at Fort Stewart, Georgia. The Naval Electronic Laboratory took care of the prime station at San Diego, California, with NRL assuming responsibility for Blossom Point and for the vehicle-tracing units on Grand Bahama Island, Antigua, and Grand Turk. In Australia an agency of that country's Joint Service Staff, the Weapons Research Establishment, built and ran the prime Minitrack station at Woomera. In South Africa, the National Telecommunications Research Center discharged these functions at the prime Minitrack station in Esselen Park near Johannesburg.[30]

Choosing sites for the Minitrack stations was also a cooperative effort. The Corps of Engineers supplied Mengel and his associates with a map study pinpointing potentially appropriate locations. The Department of State conducted negotiations for the lease of lands on foreign soil, and a Vanguard reconnaissance party under Commander Berg received substantial assistance from local representatives of the Inter-American Geodetic Survey during a tour of seventeen Latin American countries in the spring of 1956.

Using criteria previously drawn up at the Naval Laboratory, Berg's group chose six Latin American sites. They were Batista Field at Havana, Cuba; Páramo de Cotopaxi at Quito, Ecuador; Pampa de Ancón at Lima, Peru; Salar del Carmen at Antofagasta, Chile; Peldehune Military Reservation at Santiago, Chile; and Río Hata in the Republic of Panama. Shortly after the return of the Berg group in early May, NRL eliminated the Panamanian site, studies having indicated that a station at San Diego would be more useful. Indeed the day was not far off when the eyes of everyone connected with Project Vanguard would be fixed on this California station. Such was its position in the path of the critical orbit that it was always the first ground unit to receive and promulgate the glad tidings that what appeared to have been a satisfactorily launched satellite was actually in orbit.[31]

While Captain Berg's party and another Vanguard group scouted the world for real estate, tracking and telemetry specialists in Washington drew up specifications covering the material and services their electronic network would require. The Towson, Maryland, plant of Bendix Radio Division of Bendix Aviation Corporation built the Minitrack ground station assemblies, exclusive of the ground antenna arrays and the antenna feedlines. Installed for the most part in government-furnished commercial trailers, the ground stations consisted of ten major components, including rf receiver and rf power supply racks and operating consoles, rf phase measurement and phase measurement power supply racks, time standard racks precisely controlled by a crystal oscillator and analogue and digital recorder units.[32]

158

As for the Minitrack antenna arrays: In May 1956 the Laboratory authorized two companies to develop these. Both produced prototypes in keeping with Vanguard requirements, so the contract went to the Technical Appliance Corporation of Sherburne, New York, as the lower bidder.[33] Melpar, Inc., of Falls Church, Virginia, developed for the Vanguard vehicle a radar beacon (the AN/DPN–48), capable of furnishing tracking information on the vehicle during flight.[34] With minor exceptions, all of the companies involved in supplying hardware and services to Vanguard's combined radio tracking-telemetry network succeeded in meeting their by no means easy schedules. In August 1956, the finesse exhibited on NRL's side of the bargaining table moved the Chief of Naval Research, Admiral Rawson Bennett, to place on record a "comment upon the exceptionally competent manner in which Mr. J. T. Mengel . . . has conducted the pre-contract phase of the Minitrack Ground Station Units . . . ."[35]

Vanguard scientists were aware that the value of their radio tracking network would be directly proportionate to the speed with which they could convert the data acquired into usable form. For this reason they began working intensively in early fall 1955 on plans for a data-processing system.

To assist in this development Paul Herget of the Cincinnati Observatory joined the Vanguard project on a consulting basis in October. A universally respected astronomer, scholarly and firm-spoken, Herget impressed his Vanguard co-workers by the aplomb with which he performed the complex mathematical chores assigned to him. He worked closely with Joseph W. Siry of NRL, a lanky young mathematician, given to interlarding his remarks with thoughtful grunts like a doctor examining an intriguing symptom; and with tall, wispy-haired James J. Fleming, a genial data-reduction specialist with a striking resemblance to Alastair Sim of British movie fame.[36]

As the data-reduction program took form, NRL brought in two more distinguished astronomers as consultants, Gerald M. Clemence and R. L. Duncome of the United States Naval Observatory, and Hagen set up a Working Group on Orbits, consisting of Siry, as chairman, Fleming, Herget, and the two Navy astronomers. Cooperating closely with the tracking people, the group established and supervised computational procedures, prepared an ephemeris, and extracted geodetic and geophysical information from orbital data.[37]

Since the data-processing system called for the services of large-scale computers, the welcome mat was out at project headquarters when in September 1955 Cuthbert C. Hurd, then director of electronic data processing machines for International Business Machines, and other IBM representatives dropped in "to discuss the probable computer needs of the earth satellite program."[38]

In a letter thanking the IBM experts for their visit, NRL pointed out that "as now foreseen, a high-speed digital computer (similar in speed and storage capacity to the IBM 704) will be required for calculating the orbit of the satellite, [but] . . . computer plans are now in the formative stage, and it may be some time before detailed specifications can be formulated for bidding." [39]

As a matter of fact, progress toward the bidding stage was reasonably rapid. In March 1956, the Office of Naval Research invited proposals "from several possible sources for the renting of computer facilities and the furnishing of mathematical and programming services to the NRL for . . . Project Vanguard." During April, IBM and two other companies responded. IBM's bid was substantially lower than the others, and the only one that fulfilled all Vanguard requirements. At a cost of $900,000, IBM was to supply six weeks' full-time operation of its 704 computer, in addition to a number of other services at no cost whatsoever to the government. Under the free items were orbit computations during the lifetime of the satellite or the lifetime of the Minitrack, whichever was shorter, for the first three successful satellites; the services of mathematicians for coding, programming, numerical analysis and related tasks beginning on the signing of the contract; one hundred hours of computing time to check programs; the establishment in the District of Columbia of a computing center to be made available on demand; a secondary backup center to be available for emergency use within five minutes of the need; necessary communications between the primary and secondary centers; and any rehearsals necessary for working out the routine of taking Minitrack data and computing the orbit.

In June the Navy and the big business machines company entered into the requisite contract. In July NRL announced that IBM was planning to create a computing facility for Vanguard in downtown Washington. In the spring of the following year IBM's proposed remodeling of the building leased for this purpose received the blessing of the District of Columbia Fine Arts Commission, and on 30 June 1957 the attractive Vanguard Computing Center at 615 Pennsylvania Avenue, NW., opened with appropriate fanfare. Later in the summer IBM provided Project Vanguard's data processing system with standby facilities by installing a transceiver connection between the Vanguard Computing Center in Washington and the company's Research Computing Center in Poughkeepsie, New York. Although this backup arrangement would prove unnecessary, the presence in upstate New York of a 704 computer, capable of handling Vanguard data should anything go wrong with the machines in the Washington Center, was comforting to the reliability-conscious managers of the scientific satellite project. [40]

Preparation of the 704 for Vanguard use began at the Naval Research

Architect's sketch of the central IBM computing center in Washington, D.C.

Laboratory where the Working Group on Orbits determined the formulations their projected launching schedule would require. The bulk of this work fell to Herget, assisted by Peter Musen of the Cincinnati Observatory. When the Laboratory's calculations were ready, IBM mathematicians and programmers translated them into computer language. In all they wrote some 40,000 discrete operations into the machine.

Throughout most of the Vanguard launching program—from December 1957 on, to be exact—the project utilized still another high-speed computer, an IBM 709 that the Air Force had installed in the vicinity of Patrick Air Force Base on Cape Canaveral. As, during each firing, the satellite-bearing Vanguard vehicle rose from its pad at nearby AFMTC, the 709 followed its early flight with calculations based on sightings made by the big tracking radar, the AN/FPS–16. While the vehicle coasted up to the altitude where its third stage could power the payload into orbit, the preliminary information concerning the coasting phase traveled by teletype to the Computing Center in Washington.

This preliminary data indicated the speed and velocity of the vehicle at that moment. At the Washington Center technicians fed it into the 704. The machine combined it with previously prepared data on the anticipated performance of the third stage, and in this manner produced the first computations showing whether the satellite had a chance of orbiting and, if so, what the orbit would look like.

These initial computations came off the 704 while the third stage of the rocket was burning itself out above the Atlantic ten to twelve minutes after liftoff. From the Center the preliminary orbit predictions went by teletype to the Minitrack stations, enabling them to make accurate observations as the new satellite passed overhead. Each station recorded data about the satellite during the brief time it remained within range, then relayed the readings so obtained to Washington where the experts at the Center converted them to punched cards and fed the cards into the 704.

As the satellite continued to orbit the earth, the Minitrack stations reported further sightings to the Center. With these, the 704 continuously refined its preliminary orbit prediction. Eventually, usually between seven and nine hours after launching, the 704 was able to issue a definitive calculation, showing the shape of the orbit and the speed, position, and altitude of the satellite for every minute throughout the following week or ten days. Simultaneously the 704 determined the geographical locations from which the observers of the Smithsonian Astrophysical Observatory and the Project Moonwatch participants could make visual observations of the satellite.[41]

At Cape Canaveral the Vanguard technicians achieved a technological first by merging the 709 computer and the high-precision AN/FPS–16 radar into an arrangement whereby the radar fed tracking data directly into the computer and the computer, in turn, drove the plotting boards in the Central Control room on the firing range. This radar-computer-controls linkage was a case of multiple applications of the same equipment. Not only did it yield data on orbit determination for transmission to Washington, it also provided the men at the range control rack with the information they needed, during the critical early phase of a launch, to make certain that their vehicle was performing safely and according to plan.

Driven by the computer, the displays in the Central Control room gave the range safety officer a second-by-second picture of the path the vehicle was following. If the rocket swerved from its appointed trajectory and appeared likely to fall on ships or land masses below, the safety officer could press the "panic button" and destroy the flying hardware before it became a hazard. The computer-activated displays performed a similar service for the man at the third-stage firing console. A mechanism in the vehicle was set up to ignite the third stage at a precalculated point in the trajectory. Utilizing data relayed by the computer from the tracking radar, the display boards at Central Control indicated whether this mechanism was functioning properly or not. If not, the operator at the third-stage console known as "Fire-When-Ready Gridley" could push his button, thus actuating ignition of the stage from the ground.[42]

Installation of the radar-computer-controls system proceeded concomitantly with development for use at the Cape of a system for greatly accelerating the reduction to usable form of data telemetered from the Vanguard vehicle. A joint product of the Naval Laboratory and Radiation, Inc., of Melbourne, Florida, this time-saving arrangement brightened the scientific world's burgeoning list of acronyms by receiving the designation ARRF, for Automatic Recording and Reduction Facility. Housed first on a mobile trailer, later in the project's permanent hangar at the Florida missile center, ARRF provided the Vanguard scientists with final information on the performance of their birds in flight within seventy-two hours after liftoff.[43]

Creation of the project's tracking and data-processing systems required about a year and a half of planning and labor. The Minitrack network became operative in October 1957, prior to which date Vanguard relied for tracking data on existing AFMTC facilities, supplemented by a prototype of the advanced version of the simplified Minitrack station, a Mark II that NRL had established at the Cape toward the end of the preceding year. ARRF began functioning in late fall 1957 and the radar-computer-controls system began functioning in December of that year. Meanwhile other phases of the scientific satellite program had moved ahead, although in the view of hard-pressed participants, dogged by the sense of urgency implicit in their commitments, progress was never totally satisfactory.

10
EARLY TEST FIRINGS

VANGUARD launching operations were the responsibility of the Glenn L. Martin Company, and long after the project had become history, square-jawed, cigar-chewing Robert Schlechter, GLM's man in charge at the Florida missile range, was still grousing over the failure of the press to give his hands "a fair shake." It was "pretty irritating," in Schlechter's view, "to read in the newspapers such headlines as 'Devoted Navy Men Work Around the Clock at Cape Canaveral to Put Up Vanguard Vehicles.' As a matter of fact, most of those 'devoted navy men' were Martin employees." [1]

Known as the Vanguard Operations Group, or VOG, the field crew consisted of four major elements. Of these the Martin contingent was by far the largest. The others were a small group of NRL engineers, a unit charged with coordinating all phases of the field operation, and a Project Office made up of the liaison officers that the military services cooperating in the Vanguard program had assigned to the field. In addition, representatives of Martin's subcontractors and NRL's contractors joined VOG from time to time to help cope with problems connected with the services or hardware their companies or laboratories were supplying.

Headed throughout most of the program by Dan Mazur, the NRL Canaveral unit was essentially managerial. As the Laboratory's chief representative in Florida, Mazur served as VOG manager, responsible for overall technical direction of the field effort. His second in charge, with the title of test conductor, was Robert H. Gray, a slight and scholarly looking engineer who had been persuaded by Milton Rosen to leave a rocket-engine development program in private industry to participate in the satellite venture. Mazur reported directly to project director Hagen or to Paul Walsh, one of whose functions as Hagan's deputy was to keep the VOG manager abreast of Vanguard policy decisions.

As Martin's chief representative, Schlechter served as base manager, responsible for the performance of his company's field obligations. His

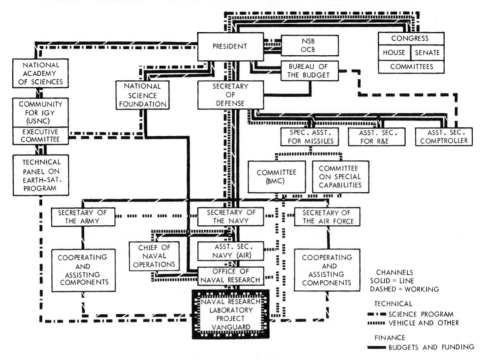

Project Vanguard's major line, policy, and working channels.

contingent consisted of a group of supervisory engineers with offices in the Vanguard hangar and two crews composed of test technicians, one at the hangar, the other at the launch complex. Before February 1958 the base manager reported directly to project engineer Markarian at the Martin plant near Baltimore; after that, to G. T. Willy, vice president and general manager of Martin's then newly created division at Cocoa, on the Florida mainland near Cape Canaveral. Schlechter's top assistant, with the title of operations manager, was Stan Welch. A West Point graduate, Welch took the edge off some of the more trying moments in the field with his running commentary on Vanguard's quaint methods. "Boy oh boy," he was fond of reiterating, "this is not the way we did things at West Point." Other top Martin men were Robert Neff as launch-complex manager, Robert Adcock and James Stoms as test conductors, Leonard Arnowitz as controls supervisor, Robert Beale as propulsion supervisor, Dave Mackey as instrumentation supervisor, and K. (Nobby) Matsuoka as mechanical engineer.

As Vanguard senior project officer in the field, Commander Harold W. (Cal) Calhoun, USN, constituted the principal link between the VOG and the missile test center command. Responsible for test control, he saw to it that the VOG carried out its field operations in a manner consistent with the capabilities and rules of AFMTC.[2]

For a brief period beginning in late spring 1956 Matsuoka of GLM was the lone occupant of the Vanguard hangar. His duties during this interval were to procure and store materiel. He also arranged for the installation of the facilities the range command had agreed to provide. In his spare time he erected a sign in the vicinity of the hangar. Reading "The Martin Vanguard Operations Group," it remained in position until it came to the attention of Commander Calhoun. Calhoun promptly replaced it with another, reading "The Navy-Martin Vanguard Operations Group."

The battle of the signs left no scars. Although along the Washington-Baltimore axis disagreements over policy and procedure continued to exercise the vocal cords of project big-shots, the Martin and NRL men sweating it out at Cape Canaveral got along reasonably well in spite of what Schlechter once described as "a rather excessive togetherness."[3]

Life at the Cape in the 1950s did not offer much in the way of distracttions or comforts. Even today's Apollo lunar launch complex at John F. Kennedy Space Center, and the vastly enlarged and more developed Eastern Test Range, successor to AFMTC, have wastelands where only the changing cloud patterns of the bright Florida sky afford some relief from the monotony of the terrain. Much of the smaller range of Vanguard days was a desolation of sand and palmetto-topped boondocks, interspersed with mosquito breeding swamps and an occasional orange or grapefruit grove.

The Air Force had provided access roads for the other projects at AFMTC, mixing gypsum powder with the sand to give them a hard surface. It did the same for Vanguard, but hurried crew members frequently preferred to make their own shortcuts, using halftracks and weapons carriers for this purpose. The practice had its risks. Before the area became government property, it had supported a few isolated farms. Forcing a path through the tangled marsh reed one day, a young NRL engineer found his pickup truck suddenly sinking beneath him, into the decaying remains of an abandoned septic tank.[4]

The hot salt breezes and mildew of the Cape created financial problems, accelerating car obsoletion and playing hob with dry-cleaning budgets. Now one of the fastest growing areas in the United States, the Cape was sparsely populated in the 1950s. Commuting distances were great and costly. Living facilities were scarce—and costly. A young bachelor attached to the VOG found himself paying $170.00 a month for a hotel room barely large enough to hold a single bed and a washstand. In the beginning Laboratory personnel assigned to the field received a $12.00 per diem.

There were groans when in April 1957 this fell to $8.00, more groans when it disappeared altogether.[5]

Danger is always present at a rocket launch complex because of the highly volatile liquid fuels, oxidizers, and acids crew members must handle. It is a tribute to the endlessly nagging safety officers of AFMTC that Vanguard accidents were minimal, with only two or three casualties so far as the record shows.[6] One of these occurred on the second-stage platform of the gantry when a workman failed to remove his arm fast enough from the interior of a vehicle afflicted with a leaking valve. Crazed by the pain from the escaping acid, he would have run off the platform and crashed on the concrete pad below had it not been for the quick thinking of John R. Zeman, the NRL engineer in charge. Grabbing the worker in time, Zeman plunged his arm into a pail of cold water. His action helped save the man's arm, but the burns were severe enough to leave permanent scars. Men working on the service tower wore terry-cloth underwear and acidproof suits. In winter, Zeman would later recall "you couldn't find one of the heavy suits, they were that popular." In summer those obliged to don them were always "looking for an excuse" to visit the "clean room" atop the gantry, the only air-conditioned spot on the pad. "Clean room" was the name given to the chambers where specialists could assemble, examine, and repair the sensitive Vanguard satellites in an atmosphere relatively free from dirt. At AFMTC the Vanguard Operations Group maintained two such rooms, one in the hangar, the other on the third-stage platform of the service tower.

When a rocket explodes in the launch-complex vicinity, there is always the possibility that the fumes released will find their way through the air-intake vents of the blockhouse, jeopardizing the lives of the men stationed in the control and instrumentation rooms—about eighty during each of the Vanguard launches. Stored in the hall of the blockhouse were piles of Scott Air Packs (gas masks). When on one occasion the vehicle did blow up on the pad, the men rushed for the packs, only to discover that nobody knew how to use them. One gathers from Kurt Stehling's lively account of the launchings that the highly trained engineers and technicians assigned to the Vanguard blockhouse never did master the Scott Air Pack. Not that the packs went to waste. Their copious folds provided useful storage space for extra cigarettes, lunch boxes, and girly magazines.[7]

As the field program gathered momentum and the stresses multiplied, VOG members invented ways of letting off steam. As soon as the road around their launch complex was completed, they began staging drag races. The range police objected, citing the danger to life and limb. The Vanguard crew retorted that they had a right to do what they pleased with their own lives and limbs, and went on racing. Dan Mazur contributed

168

to the merriment with his by no means infrequent wails that someday the range safety officers might find it necessary to command-destruct one of his precious Vanguard vehicles in the air before it could complete its appointed course. So feelingly expressed were Mazur's fears on this score that one day some members of his crew packaged up a piece of battered Vanguard hardware and sent it to friends in Germany. With it went appropriate instructions. These the Americans abroad conscientiously fulfilled, with the result that in time the damaged steel returned to AFMTC, addressed to base-commander Yates. Lettered on the steel, as the general discovered on opening the package, was a message reading, "Attention, Mazur: What's the big idea, impacting your damned hardware on German soil!" [8]

Each of the fourteen Vanguard launchings raised particular problems. In every case, however, the procedures leading up to and during the launch of the complete vehicle were roughly identical. The major subcontracted elements—notably the engine for the first stage, the second-stage power plant, and the third-stage solid-propellant rocket—received acceptance tests at their points of origin before moving on to the big Martin plant at Middle River, Maryland. There men working in the Vanguard shop area "married" the components of each stage, insofar as was necessary, and installed instrumentation. Then followed a series of systems tests, after which the plant crew assembled and tested the entire vehicle in a "silo" or tower built for this purpose on the plant grounds. None of these tests at the Martin plant was a "hot" or "static" test; none, in other words, necessitated the firing of the rocket engines, a much too hazardous operation for a plant located in a heavily populated area. Working out of an office at the Martin plant, James M. Bridger, director of the Vanguard vehicles branch, functioned as project engineer for the Naval Research Laboratory. He, Walsh, Berg, and other NRL experts monitored the proceedings at the factory and took delivery of the vehicle, subject to the approval of Hagen and his technical director, Milton Rosen. Following NRL acceptance, plant workers disassembled the vehicle and shipped it south, each stage traveling on a specially built trailer.[9]

At AFMTC the field crew put the vehicle through further inspections and tests, first in the hangar and then at the Vanguard launch complex, pad 18A. At the pad all first-stage rockets underwent static tests. A static test can be defined as a flight firing of a liquid-propellant rocket without flight. With heavy bolts holding the rocket to the launch stand, the crew ignited the first-stage engine and permitted it to fire for a specified period.

No Vanguard rocket ever got away during these hot runs, but such things have happened. Jim Bridger still enjoys recalling a static at White Sands in the early 1950s when a portion of one of the Viking rockets broke loose. To Bridger, the sight of the escaped hardware soaring to an altitude of 17,000 feet was "startling." The spectacle put Bridger in mind of Maurice Maeterlinck's hilarious description, in one of his nature essays, of the flight of the bumblebee.[10]

The primary purpose in static firing the Vanguard first stages was to make certain that the engine, previously tested at Martin, was properly mated to the rest of the rocket, and that propulsion systems, instrumentation, stabilization systems, and controls were in working order. As for the second-stage rocket, the question of the extent to which that too should undergo statics evoked considerable debate. Everlastingly concerned with reliability, NRL's vehicle experts favored a fairly extensive use of such tests as a means of verifying the satisfactory behavior of the second-stage propulsion system. Martin experts conceded the need for such data but wanted the tests held to a minimum to save time. Over the long pull the company's position prevailed. Only three stages underwent static firings in the field. Statics performed in connection with TV–3 and its backup vehicle, TV–3BU, yielded useful information, but a static firing in the fall of 1958 of the second stage of one of the mission vehicles, SLV–3, damaged the rocket. As a result, the Vanguard management abandoned the practice for the remainder of the program.[11]

The first-stage statics brought the prelaunch operations at the pad more or less to the half-way mark. Next came alignment checks, instrumentation calibrations, and system functional tests, culminating in the vertical functional test. Conducted with full range support, this test was in effect a dry run of the forthcoming countdown and flight. A limited version of it, the flight readiness test, completed prelaunch operations.

Preparations for the flight itself usually began two days or more before T−O (takeoff). At the AFMTC solid-propellant storage area, members of the VOG assembled and resistance-checked the third-stage motor and other ordnance items. They then transported these to the pad and installed them in the erected vehicle. Other preparations on the day before flight included checks on the satellite, the vehicle propulsion system pressures, the pipelines supplying water to the launch stand, and the fire-fighting facilities.

On flight day, operations began approximately eighteen hours in advance of launch with checks to ensure that all vehicle systems were in order and the first two stages were ready to receive propellants. About eight hours before launch, technicians installed the satellite on the third stage of the vehicle.

Those responsible for preparing the last phase of the preparatory sequence—the countdown—normally wrote into it a one-hour planned hold. Scientific considerations prompted the practice. From TV–3 on, the

Martin Co. and Aerojet General Corp. personnel wear protective clothing while fueling the Vanguard second stage with white inhibited fuming nitric acid.

Martin Co. personnel at Baltimore assemble the guidance mechanism in the second stage of Vanguard.

The nosecone of Vanguard is fitted into position.

main objective of every Vanguard launch was to place an experiment-bearing satellite in orbit. To render it possible for the experiments to acquire the data desired, it was often important that the satellite enter orbit under certain circumstances having to do with the position of the earth relative to the lunar system and other variables. The built-in hold increased the likelihood of the satellite achieving orbit under these previously calculated optimum conditions by providing the launching crew with extra time in which to cope with unforseeable delays. If the countdown proceeded perfectly, the crew did nothing during the planned hold. If forced holds carried the procedure beyond a previously computed time point, the crew had no choice but to scrub the launch and start all over again.

Ordinarily the countdown began five hours before launch.[12] At $T-255$ minutes technicians turned on the satellite and checked it. At $T-95$ minutes liquid oxygen (lox) began pouring into the oxidizer tanks of the vehicle. At $T-65$ minutes the gantry crane retired from the flight firing structure. At $T-3$ minutes the time-unit used for the countdown changed to seconds, and instrumentation men shifted the telemetry, radar beacons, and command receivers to internal power. At $T-30$ seconds the cooling-air umbilical dropped and the lox-vents on the vehicle closed. At $T-0$ the fire switch closed, the electrical umbilical dropped from the vehicle, and about six seconds later $(T+6)$, if all was well, the vehicle lifted off.[13]

During peak periods the VOG ranged from one hundred to one hundred fifty men. In October 1956, about fifty were working in the project's temporary assembly building, Hangar C, or at its still unfinished launch complex when Viking 13, refurbished and renamed Vanguard Test Vehicle Zero, or TV-0, arrived at the hangar. A month later crewmen had transported it to pad 18A and were erecting it on the old Viking launch stand recently shipped from White Sands for use at the Cape pending arrival of the more advanced flight firing structure Martin had designed and Loewy Hydropress was fabricating for the Vanguard program.

TV-0 consisted of only one stage. Flight testing of a full-fledged three-stage Vanguard vehicle lay in the future. The project managers had reasons for initiating their launching program on this modest level. It was important that before attempting to fly the entire vehicle they familiarize themselves with the operations and the range safety and tracking systems at AFMTC.[14]

Rain was falling when an hour after midnight, 8 December 1956, the countdown reached its final seconds. A variety of difficulties had plagued the final launching procedures. Snarls at the range telemetry building and at Central Control had necessitated two holds, the appearance of a ship in the waters of the impact area, another. Nerves were jumping in the

TV–0 on the launch stand at Cape Canaveral, launched 8 December 1956.

crowded control room of the blockhouse, with Colonel Gibbs, the Air Force's conscientious project officer, shouting dire predictions at Bob Schlechter, the man in charge. "It's gonna blow up, Bob," Gibbs kept insisting. "Cancel! It'll never fly!"

But it did fly. Lifting off at 1:05 a.m., TV–0 achieved an altitude of 126.5 miles and a range of 97.6 miles. One of the objectives of the launch was to test Vanguard's newly developed Minitrack transmitter. With this in mind Mengel's tracking team had devised and Martin had installed in the vehicle a special Minitrack package. At T+120 seconds, two minutes after launch, the triggering device of the package—a timer—powered two bellows-contained squibs, causing them to ignite and expand, thereby withdrawing a releasing key and allowing a compressed spring to extend and eject a small sphere equipped with "roll-up" antennas and enclosing a Minitrack transmitter. Without difficulty the ground receiving units at AFMTC, the Laboratory's Mark II tracking station among them, picked up the little oscillator's plaintive beep as the ejected package descended into the sea.[15]

In mid-December a conference room at AFMTC headquarters was the scene of a post-mortem on the first Vanguard flight test. Of the thirty-two men in attendance, twenty-five were members of the base command or Pan Am and RCA technicians involved in the intricate range-support activities connected with the satellite program. On hand for the Naval Laboratory, in addition to VOG chieftains Mazur and Gray, were Joseph Siry, head of the Vanguard theory and analysis branch, and his handsome, blue-eyed assistant, Richard L. Snodgrass. Martin's representatives were Schlechter and dark, stocky, thoughtful Joseph E. Burghardt, the company's assistant project engineer for aerodynamics and propulsion, who, although stationed at the Middle River plant, was a frequent visitor to the field.

Facts brought out in a lengthy briefing—all verified by subsequent analysis—showed that on the whole the TV–0 launching had achieved its prescribed objectives. During powered flight of the vehicle, the performance of all components had been "either satisfactory or superior." Rocketborne instrumentation and telemetry systems had functioned "excellently," ground instrumentation coverage had been "adequate."

Back in Washington, however, expressions of pleasure in these results were muted. Concern over the general status of the program dominated the discussion. Several of Martin's subcontractors were finding it impossible to meet their delivery schedules. Because of this and slippages in other aspects of the undertaking, all of the firing dates previously established for 1957 had already been substantially advanced. Now little hope remained that even these frequently rescheduled dates could be realized.[16]

The plodding progress of the next few months added to a mounting sense of frustration. At the range, the outstanding events of January and February 1957 were the arrival at the hangar of the second Vanguard test vehicle, TV–1, and the completion of all of the project's permanent field facilities with the exception of hangar S.

TV–1 was a two-stage vehicle. Its booster was the last of the Viking research rockets, No. 14, slightly modified for Vanguard purposes. A product of the Grand Central Rocket Company, the second stage was a prototype of the solid-propellant rocket destined to become the third stage of the finished Vanguard vehicle.

174

Successful launch of TV–1, 1 May 1957.

Attending a Project Vanguard staff meeting in the director's office at NRL: left to right, Daniel G. Mazur, Manager, Vanguard Operations Group, Patrick AFB; James M. Bridger, Head, Vehicle Branch; and Commander Winfred E. Berg, Navy Program Officer.

Although the difficulties encountered during the prelaunch procedures at the hangar and on the pad were comparatively minor, their correction ate up precious time. Hopes for a February flight vanished rapidly. It was late March before the crew was able to erect TV–1 on the old Viking stand at the launch-complex. In early April static tests began, and in the dark hours of 1 May 1957—at 1:29 a.m.—the second Vanguard test vehicle lifted off.

As set forth in the test plan, the primary purpose of the launch was to flight-test the third-stage prototype for spin-up, separation, ignition, and propulsion and trajectory performance. A secondary objective was to further evaluate ground handling procedures, techniques and equipment, and the in-flight vehicle instrumentation and equipment. Studies of the telemetered data acquired during flight would show that all objectives were met. The first-stage rocket performed "about as expected." The second stage (actually the Vanguard third stage) separated and fired "nearly as expected" with a total burning time of about thirty-two seconds. It was this satisfactory first firing of the third stage that prompted NRL, during the following July, to inform the Martin Company that from TV–3 on all Vanguard vehicles were to possess satellite-bearing capacities.[17]

In the world of the mid-1950s two successful rocket launchings in a row added up to a singular accomplishment. As TV–1 roared to an altitude of 121 miles, Dave Mackey of GLM, unofficial comic of the blockhouse gang, voiced a common sentiment. "I wonder," he mused, "if success will spoil Project Vanguard?"[18]

Less than a month later, the members of the VOG in general, and their ebullient boss, Dan Mazur, in particular, were telling themselves that Project Vanguard had become Project Impossible. Getting the project's third test vehicle, TV–2, out of the Martin plant, down to the field, onto the launch stand, and up in the air was an ordeal of more than five months' duration. So many troubles beset the process that at one point Mazur would have resigned in disgust had it not been for the gentle-spoken persuasiveness of project director Hagen.[19]

TV–2 had the external configuration of a complete Vanguard vehicle, although strictly speaking it was not complete. All three Vanguard stages were there, but only the first stage, consisting of the Martin tankage and the General Electric X–405 liquid-propellant engine, was live. The second and third stages were inert dummies.[20]

Today former Vanguard men can say calmly that the nightmare of TV–2 was "just one of those things." Back in the Vanguard days, Jim Bridger has commented,

> we were aware that the ultimate source of our funds, the Department of Defense, had reservations about the value of a purely scientific missile development. Consequently we made political fodder out of saying the Vanguard vehicle was just an outgrowth of the Viking research rocket. Frankly, that was an exaggeration. We did indeed bring Viking experience to the Vanguard program and the first-stage engine was a take-off, albeit a complex one, from General Electric's Hermes A–3B engine; but for all practical purposes the Vanguard vehicle was new, new from stem to stern. More to the point, it was an awfully high-state-of-the-art vehicle, especially the second-stage rocket. In the nature of things the business of developing the vehicle and getting the bugs out so it would work was fraught with difficulties.[21]

This fact, obvious as it would become in retrospect, was of no consolation to the harried men who in the summer of 1957 began the long struggle to get the bugs out of TV–2.

The extensiveness of these bugs came to light early in the summer during the vertical interference and acceptance tests of the vehicle at the Martin plant. Some of the structural discrepancies uncovered at that time gave only minimal trouble, the company coming up quickly with remedies satisfactory to NRL. More serious was the failure of the roll jet and pressurization systems to perform in accordance with specifications. To some extent these had to be redesigned. Since this was a time-consuming job and time was of the essence, Martin asked the Laboratory for permission to ship TV–2 to Cape Canaveral where the field crew could begin receiving inspections in the hangar while GLM redeveloped the faulty systems.

Reluctantly the Laboratory acceded to this suggestion. In a stern letter to the contractor, Hagen pointed out that although Martin's proposal for sending to the field "an unaccepted, incompletely developed vehicle" violated "sound principles of operation, the Laboratory agrees that this is the only way to have at least some chance of maintaining the firing schedule." The Laboratory, therefore, "will provisionally accept TV–2" with the understanding that in the near future "Martin shall qualify and deliver all outstanding components of the roll jet and pressurization systems." [22]

Hagen's letter got TV–2 out of the Martin plant. It also placed additional burdens on the field crew. When the redeveloped systems were ready, members of the VOG had to install them, a procedure normally carried out at the factory. By this time—late July—NRL and Martin had concluded that all along the line many modifications of the vehicle were going to have to be made in the field instead of at the factory as originally planned. Given the time pressures on the program, no other arrangement was possible, but it did not make the field workers happy. Frequently the required modifications were so basic as to amount to design changes. Taking care of these at AFMTC was difficult since more often than not the necessary tools and spare parts were unavailable there and had to be improvised on the spot or procured from distant points. [23]

With the arrival of TV–2 at the Cape in early June, new troubles presented themselves. Profound groans and profane gripes filled the Vanguard hangar as inspection revealed that both the first-stage tankage and engine contained "fine filings, metal chips and dirt." The VOG crewmen could clean the tankage, but getting the dirt out of the engine was beyond their capacities. Back went the motor to the General Electric plant at Malta, New York, with orders for its makers to send another to the field. In July Rear Admiral Rawson Bennett, Chief of Naval Research, covered the situation in one of his always admirably dispassionate reports to the Chief of Naval Operations. The presence of "extraneous material" in the motor, the admiral wrote, along with the delay "occasioned by repairs to damaged items, the clean-up procedure, . . . and now the installation of a new motor makes it appear that the earliest possible flight firing will be the last week in August"—a statement that piled optimism on euphemism. [24]

August passed with a "possible flight firing" seemingly as remote as ever. At pad 18A the only encouraging sign was the disappearance of the old Viking launch stand and its replacement by the Vanguard static and flight firing structure. Even this was not for keeps. Some of the Martin Company specialists were fearful that under some circumstances the gimbaled engine of the first stage might not clear the fixed opening in its stand during liftoff. Already they were working on designs for a stand with movable components, capable of springing away automatically as the vehicle rose. There was disagreement at the company as to the necessity for this change, but those in favor won the argument. Eventually the Martin-designed retractable or breakaway firing stand would find its way to Cape Canaveral, to become one of Project Vanguard's several contributions to the advancement of missilery. [25]

Second stage of Vanguard being hoisted into position.

The replacement engine ordered from General Electric arrived at the hangar in good time. Not until 22 August, however, did the prelaunch preparations reach the point where the crew at the pad could attempt a static firing.

A static test, like a flight test, involves a lengthy countdown. The 22 August one began on schedule, but at $T-290$ minutes accumulated difficulties forced Mazur and Schlechter to call a hold that lasted for more than five hours. Soon after resumption of the countdown, new difficulties arose. During the first attempt to pressurize the fuel tanks, a lox vent failed to relieve excessive pressure. When the vent refused to close fully during several succeeding attempts, the VOG bosses did the only thing they could. They scrubbed the test and instituted an investigation. The presence of water in the lox vent indicated that freezing had prevented it from closing. During all future launching operations, as a result of this discovery, the crew subjected the lox system to a constant nitrogen purge from the start of the countdown until the point at which lox servicing began.[26]

For Mazur the first attempt to static fire TV–2 was a domestic as well as a professional disaster. During the test, his wife, who had not yet joined him at the Cape as she would later, sent a telegram, informing him that she and two of their three children had contracted the mumps and needed daddy at home. Mazur stayed with his job. Mrs. Mazur, he would reveal later, "never forgave me. To this day, whenever we have an argument, she reminds me how back in the summer of 1957, I let her down in her hour of need." At some point during that trying summer, Admiral Bennett paid the field crew a visit. Closeting himself with Mazur and Schlechter, he demanded, "what's going wrong down here anyhow?" Mazur's reply was, "Just one thing: Instead of rockets, Martin is sending us garbage"— only, according to Schlechter, Mazur's final word was shorter and more colorful. Later, in a more relaxed mood, the VOG boss snapped off to his friend Schlechter a teletype reading:

> Rockets are large, rockets are small,
> If U get a good one, give us a call.[27]

The project bosses at the Naval Laboratory in Washington shared Mazur's chagrin at the situation. As Vanguard technical head, Rosen found only occasional fault with Martin's design work on the vehicle. On the whole he regarded it as excellent, but where the company's shopwork on the vehicle was concerned he deplored what seemed to him at times to be a carelessness bordering on indifference. Repeatedly he urged project director Hagen to "crack down" on the company. The frequent disputes between his staff and the contractor had convinced him that he was dealing with "two tigers." In the interests of keeping the program moving, he hesitated to take a step likely to exacerbate existing differences.[28]

The difficulties with TV–2, however, were too much even for the judicious and even-tempered project chief. As complaints continued pouring in from the field, he got off a sharply worded remonstrance to GLM. "The performance of the Martin Company in regard to TV–2," he wrote, "has been unsatisfactory and increases the laboratory's concern about the ability of the contractor to meet launch schedules in the future. Specific items have been discussed in detail during conferences and will be further stated in writing if the contractor so desires. The contractor is urged to bend every effort toward maintaining or bettering the present launch schedule." [29]

Hagen's reprimand failed to alter the course of events with respect to TV–2. Its long-range effect on the Vanguard program, however, was a salutary one. Hardware difficulties would continue to arise, but in the future the source of few of them would be in the Martin plant. It is of interest to add in connection with this aspect of the program that some members of the Martin company's Vanguard group have criticized their top management's handling of the satellite project. In the beginning, to quote one of them, "the Martin managers didn't ride herd on the Vanguard job as vigorously as they should have. In a job of this sort the managers of the company should walk the floor. The Martin managers failed to do that at first; when later on, they did so at least to some extent, things improved immensely. The three most successful Vanguard launches were all preceded by a tightening up of procedures and a greater watchfulness on the part of the management." According to this same critic, "another mistake" of the GLM officials was their failure "to get into bed with the customer. The Martin people did a swell job, but somehow the Martin managers were never able to convince the Naval Research Laboratory that they had. All along the line there was an unfortunate breakdown in communications between company management and customer." Evidence that in the beginning at least the Martin managers regarded the scientific satellite program as though it were a "poor relation" is provided by what has come to be known as "the era of the bird-droppings." For several months the company installed the designers of the Vanguard vehicle in the upper reaches of an old plant where broken windows provided convenient passageway for the sparrows living in the girders. Drawings left open on a draftsman's table at night were seldom quite the same by the following morning.[30]

The second TV–2 static test, attempted four days after the first, encountered even worse luck. Among other things, the blast deflector tube of the firing structure suffered serious damage. During the helium pressure tests, excessive leakage showed up in the turbine and deflector-plate seals of the engine. Again the crew thought it best to remove this component and ship it back to General Electric. By this time, fortunately, TV–2's backup vehicle TV–2BU, had arrived, so a spare motor was available. The crew installed it, and grimly prepared for a third attempt on 3 September. That, too, had to be scrubbed when at $T-245$ minutes the main pressurization system regulator exhibited behavior characteristic of a dangerously dirty valve.[31]

September saw three static-test attempts in all—and three heartbreaking scrubs. October was well underway before static-test number seven satisfied the VOG bosses that TV–2 was ready for launching. Two flight firing attempts during the second half of the month had to be called off long before the completion of countdown. The third was a resounding success: with a long succession of difficulties now overcome the first flight to be attempted with the Vanguard external configuration carried a 4,000-pound payload to an altitude of 109 miles and to a downrange distance of 335 miles as planned. All test objectives were realized. Performance of all components was "superior." The flight showed that the Vanguard first stage operated "properly at altitude," that "conditions were favorable for successful separation of the first and second stages," that launch-stand clearance in low surface winds was "no problem," and that "there was structural integrity throughout flight." The test also demonstrated the existence of "dynamic compatibility" between the control system of the vehicle and the structure.[32]

At the time of the flight, however, there was little rejoicing in the Vanguard blockhouse. Relief was the prevailing sentiment there when the word came that the vehicle had completed its appointed course and fallen into the ocean. According to Kurt Stehling,[33] the unspoken thought of the men who had carried TV–2 through its many trials and tribulations was, "Let the fish have it."

They got it on 23 October 1957. By that date, drastic changes had overtaken Project Vanguard. Some reflected policy decisions within the project itself. Others were the outgrowth of that turning point of the Space Age, the launching into orbit by the Soviet Union of the first man-made earth satellite, *Sputnik I.*

Launch of the difficult TV–2, 23 October 1957. This was the first three-stage configuration of Vanguard, although the upper two stages were inert.

FROM SPUTNIK I TO TV-3

THE VANGUARD field crew was still struggling at Cape Canaveral to put up TV–2, its third test vehicle—the one designed to test the first stage—when on Friday, 4 October 1957, the news broke that *Sputnik I,* a 184-pound sphere had been launched about 5:30 p.m. that day by the Soviet Union and was circling the earth.

Earlier in the week, on Monday, 30 September, scientists representing the Soviet Union, the United States, and five other nations had assembled at the National Academy of Sciences in Washington, D.C., for a six-day CSAGI conference on the rocket and satellite activities of the International Geophysical Year. A speaker at the opening session was Sergei M. Poloskov, member of the Soviet delegation. Poloskov's subject was "Sputnik," the Russians' word for "traveling companion" and the name they had chosen for the satellite they were preparing to launch. The U.S.S.R. had long since served notice of its intent to develop a satellite-launching program as one of its contributions to the IGY. Nevertheless, there was a stir among Poloskov's listeners when he used an expression that could be literally translated as "now, on the *eve* of the first artificial earth satellite." There was another stir when he revealed that the transmitters in the projected Soviet satellite would broadcast alternately on frequencies of 20 and 40 megacycles. In 1956, CSAGI, the international ruling body for the IGY, had adopted a resolution stipulating a frequency of 108 mc as standard for all IGY satellites. Speaking for the United States at the CSAGI session in Washington, Homer Newell pointed out to the Russian scientist that Project Vanguard's radio tracking stations were set up to receive signals on the IGY established frequency. Since adapting the American Minitrack to receive the lower Soviet signals would require time and money, he asked Poloskov to specify when his country hoped to put its first satellite in orbit. The deftness with which Poloskov sidestepped Newell's question, along with similar questions from other delegates, produced a roar of laughter in which the Russian scientist himself finally joined. All he would say was that when the Soviet satellite materialized, he hoped the Vanguard tracking stations would collect the data it transmitted and send them to Moscow.[1]

On the following Friday evening the delegates to the conference were guests of a reception in the ballroom on the second floor of the Soviet embassy. Among the reporters on hand was Walter Sullivan of the *New York Times*. When shortly after 6 p.m., Sullivan received a phone call from his Washington editor, he made a point of getting as quickly as possible to Richard Porter, member of the American IGY committee and chairman of its technical panel. "It's up!" he whispered. Although Porter had been convinced for days that a Soviet launching was "indeed imminent," his normally red face was redder than usual as he and Sullivan wedged through the crowd in the embassy ballroom to relay the news to Lloyd Berkner, this country's official delegate to CSAGI. Berkner clapped his hands for silence. "I wish to make an announcement," he said. "I've just been informed by the *New York Times* that a Russian satellite is in orbit at an elevation of 900 kilometers. I wish to congratulate our Soviet colleagues on their achievement." [2]

It was a gracious and dignified beginning to a period of mental turmoil and vocal soul-searching in the United States that can scarcely be described as dignified. In retrospect it is easy to smile at some of the exaggerated alarms and groundless assumptions that filled newspaper columns and trumpeted from public platforms as the significance of the Soviet feat became apparent. The smug chuckle of hindsight, however, cannot efface either the importance of the event or the intensity of the change it wrought in American thinking. Girdling the earth once every 96.17 minutes, the first Russian satellite—later referred to as *Sputnik I* to distinguish it from its successors—was a sphere approximately twenty-two inches in diameter, made of aluminum alloys and equipped with four spring-loaded whip antennas. It carried two continuously signaling transmitters and an instrumentation package primarily designed to disclose the effects of meteoritic collision. The power supply for telemetering information to ground stations was a chemical battery. The perigee of the initial orbit was 142 miles, the apogee 588 miles. The inclination to the equator was 64.3°; speed at perigee was 18,000 miles an hour, and at apogee, 16,200 miles an hour. The satellite itself would fall from orbit on 4 January 1958. Its two transmitters would fail twenty-three days after launch—but their arrogant beep-beep would continue to sound in the American memory for years to come. "Sputnik night," as the night of 4–5 October 1957 came to be called, was an historic watershed. Almost immediately two new phrases entered the language—"pre-Sputnik" and "post-Sputnik." In England the London *Daily Mirror* proclaimed the birth of the "Space Age" in huge headlines, and changed its slogan to claim, not the "biggest daily sale in the world" but the "biggest . . . in the UNIVERSE." Gone forever in this country was the myth of American superiority in all things technical and scientific. The Russian success alerted the American public to deficiencies in their school system, to the need for providing their young people with an educational base wide enough to permit them to cope with the multiplying problems of swift technological change.[3]

American response to the Russian triumph varied considerably, depending on its source. The alarm exhibited by large sections of the public did not materialize immediately. In New York City, on "Sputnik night," phone calls poured into the offices of the Hayden Planetarium and the American Museum of Natural History. Practically all were from people seeking more information than the Soviet bulletin to the American press had provided—mostly amateur astronomers and ham radio operators eager to get down to the happy business of trying to acquire and track the world's first man-made satellite. At central police headquarters, a spokesman at the big switchboard, the activity of which is regarded as an index to public anxiety, reported no inquiries whatsoever. On the following day a *Newsweek* correspondent in Boston wrote that the "general reaction here indicates massive indifference." From Denver another *Newsweek* writer wired his home office that there "is a vague feeling that we have stepped into a new era, but people aren't discussing it the way they are football and the Asiatic flu." Before a week had gone by, New York's silence, Boston's "massive indifference," and Denver's bewilderment had melted away before a mounting and all but universal furor. To the majority of Americans the Soviet feat came as a total surprise. It needn't have, according to some of the commentators participating in the storm of charge and counter-charge that followed. For some time the United States government had been in possession of intelligence reports showing that Russian missilry was well advanced and that the U.S.S.R. had hardware capable of placing a satellite in orbit.[4] In the confused post-Sputnik days, science reporters and others contended that if the Administration had made its knowledge public, the launching of the Soviet satellite would have had a less traumatic effect on the American people. Perhaps so, perhaps not. In the halcyon pre-Sputnik days the American people would probably have paid little more attention to such information, its official source notwithstanding, than they had paid to already existing evidences that Soviet science was developing at a phenomenal rate. Most Americans were aware that Russia had created an atom bomb more quickly than American authorities had considered likely. They knew that Soviet work on the hydrogen bomb had kept pace with that of the United States. As recently as August 1957, the U.S.S.R. had claimed a successful intercontinental ballistic missile

Cartoon reaction to Sputnik I.
(Courtesy Thomas Flannery,
Baltimore Sun.)

test. None of these facts, however, had registered deeply in this country. Nor had the occasional story in the press hinting at an upcoming space breakthrough by the Soviet Union. When Sputnik appeared, the reaction of the public, taken as a whole, was a compound of awe, surprise, chagrin, and fear. The Russians had beaten us into space! More to point, if they could put up a harmless scientific satellite—assuming it *was* harmless— what was to prevent them in the near future from putting up a larger one equipped with nuclear warheads! To be sure, not all was gloom and worry. The Americans invented Sputnik jokes, and laughed at jibes originating overseas. Bars around the country advertised "Sputnik cocktails," one third vodka, two thirds sour grapes. In Poland the people quipped that at last Russia had a smaller satellite than Albania, and in West Germany they coined a new name for America's still unorbited Vanguard. They called it "Spaetnik," *spaet* being the German word for late. After Russia had launched a second and even more spectacular satellite, a reporter's query to Nikolai A. Bulganin, the Soviet premier—"When are you putting up the third one?"—brought the grinning riposte, "It's America's turn now." [5]

Official reaction to *Sputnik I*—which is to say, high-level government reaction in Washington—was also marked by surprise, but of a different sort. The government was startled by what Defense Secretary Wilson described as the public's "jitters." At an October press conference and in two subsequent television appearances, President Eisenhower undertook to reassure an agitated nation. The chief executive conceded that the Soviet achievement was a "political defeat" for the United States. He stressed, however, that this country and the U.S.S.R. were not engaged in a space race, a statement that the Russian leaders, with the same indifference of officialdom to the facts of life, had also made. Granting the remote military potentialities of Sputnik, the President asserted that it "does not raise my apprehensions . . . one iota" about the national security.[8]

Political reaction, emanating principally from Congressmen and state governors, ploughed a familiar furrow. Most of it was not so much concerned with what the Russians had done as with what the Americans had so far failed to do. Why was this country behind in the space race? Who was to blame? As always in cases of national distress, the White House headed the list of targets. Spokesmen from both major parties accused the Eisenhower Administration of "penny-pinching," "complacency," "lack of vision," and "incredible stupidity" where both the American missile program and Project Vanguard were concerned. They reiterated the President's frequently quoted description of scientists as "just another pressure group" and cited Secretary Wilson's confessed indifference to basic research and the avowed indifference of presidential aide Sherman Adams to "an outer space basketball game," a heavy-handed reference to the small Vanguard satellite. Former President Truman mirrored a segment of public opinion when he attributed the space-lag to the "[Senator Joseph] McCarthy era." He charged that "official persecution" of prominent scientists in the early 1950s had deprived America's missile and satellite programs of some of the country's "best brains," a statement scarcely calculated to flatter the highly capable scientists working with Project Vanguard.

American scientists, including those associated with Vanguard, had for some time been aware that the U.S.S.R. possessed the capacity to launch a satellite. It would be an exaggeration, however, to say that all of them had faced up to the full implications of their own knowledge. Even scientists are not immune to wishful thinking. Before Sputnik some of those in America had been as willing as the general public to discount Soviet claims. After Sputnik many were quick to praise. Joseph Kaplan, chairman of the American IGY committee, pronounced the Soviet launch "really fantastic." Kaplan pointed out that the Russians obviously had developed a launching vehicle of tremendous thrust, since the sphere they had orbited was eight times heavier than the larger of the two spheres then available to the American satellite program. A further indication of the power of the Soviet launching vehicle lay in the orbit the Russians had used because of the more northern latitude of their launch site. The Vanguard managers were planning to take advantage of the earth's rotational velocity, about a thousand miles an hour at the equator, by putting their satellite into a west-to-east orbit, between 30° north and south of the equator. Russia was denied much of this advantage. Circling roughly from 65° north to 65° south latitude, Sputnik's orbit was more north-south than east-west.

In the absence of an official explanation from the U.S.S.R., American scientists offered varying theories as to why the Russians had used 20-mc and 40-mc frequencies in their payload instead of the 108-mc frequency prescribed by CSAGI and used by the United States. One theory was that the Russians saw a propaganda value in using the 20- and 40-mc frequencies. These could be picked up by the sets of amateur radio operators around the world whereas the higher frequency required more sophisticated receiving equipment. Another theory held that the Russians simply did not have receivers capable of picking up signals at the higher frequency. A third theory, easily the most convincing in view of the advanced state of Soviet science and technology, was that the lower frequencies were more suitable to the scientific objectives of the Soviet launch and to the relatively low altitudes of Sputnik's orbit. Another subject of speculation among American scientists was whether or not the Russians were trying to follow their satellites with cameras. A few months prior to the launching of *Sputnik I,* a group of Soviet scientists participated in a symposium on cosmical gas dynamics at the Smithsonian Astrophysical Laboratory in Cambridge. Two members of the group, both famous astronomers, talked at length with Fred Whipple, director of the Observatory and head of the Vanguard optical-tracking program. Although the Russian astronomers were noncommittal as to their country's plans, Whipple got the impression that the U.S.S.R. had already developed some sort of optical-tracking system in connection with its satellite program.[7]

As for the reaction of the Vanguard managers to *Sputnik I,* the state of their feelings can be left to the imagination. It was one thing to have surmised, as many of them had, that Russia was on the verge of orbiting a satellite. It was another to realize that the satellite was actually up there. Lured before the television cameras of a news program on the evening of 4 October, before he could put his thoughts in order, Admiral Rawson Bennett dismissed Sputnik as "a hunk of iron almost anybody could launch." The statement was a tribute to the naval research chief's loyalty to Project Vanguard but as an assessment of the situation, to quote the Russian chairman Nikita Khrushchev, it was a "cosmic boo-boo." [8]

Russia's use in her satellite of lower broadcasting frequencies than those of the United States presented the Vanguard radio-tracking people

Minitrack antenna arrays.

with a severe problem. Soon after the news of Sputnik reached Washington, Mengel and his aides, accompanied by Joe Siry and other orbit-computation experts, were on their way to the Vanguard control room on the second floor of building 72 at the Naval Research Laboratory. Most of them would remain there around the clock for the next three days. The six prime Minitrack stations along the Vanguard "fence" were ready—ready, that is, to receive signals from a satellite transmitting on the IGY-established frequency of 108 mc. The job of the tracking technicians was to convert them to the Sputnik frequencies fast enough to enable the big computing machines in Washington to calculate and predict the course of the Soviet satellite. Toward the end of the summer building crews had completed construction work on the six major Minitrack stations, the men selected to operate them had reported for duty, and the tracking-system heads were laying plans for a series of dry runs to test the network. The dry runs never took place. Sputnik arrived before they could start, and when three weeks later the Russian satellite ceased transmitting, the network had become operational, tracking Sputnik and relaying its position to Washington. By that time, as an Army engineer remarked later, the U.S.S.R. had provided Minitrack with "the wettest dry run in history." [9]

One of the difficulties created by the launching of Sputnik was that by fall 1957 the Signal Corps had not yet completed the communications network it was in the process of establishing between Vanguard headquarters and all Minitrack units. Direct communication from the Washington area reached only to the stations at Fort Stewart, Georgia, and Antofagasta, Chile. To reach the other prime stations, headquarters had to resort to slower means—commercial cable, the Inter-American Geodetic Survey radio net, amateur radio, and long-distance telephone. Over whatever means were available, messages went quickly from the control room at NRL, providing station crews with such technical information as the men at NRL could collect, and suggesting ways for putting emergency equipment to work and for modifying antenna arrays. In addition, NRL air-shipped special receivers to all stations.

No one could take seriously the naïve statement of one bitter but ill-informed observer that "Sputnik caught Project Vanguard with its antennas down." [10] Certainly they did not stay down long. In mid-October C. B. Cunningham, senior NRL scientist for the station at Lima, Peru, was able to send his superiors a glowing report.[11] "Our first track of the USSR satellite," he wrote—

occurred on Friday morning, 11 October. It was a thrill for all hands. Since that time we have tracked every passage but one. In accordance with the NRL request and instructions, we constructed two dipole antennas, tuned to 40 megacycles, which we mounted 34 inches above the 108 mc dipoles, and at the center of the 108 mc array. We used 2×4 end supports and lacing cord. We did have to improvise for coaxial panel connectors and T connectors. We just didn't have any, so we got along without them. Anyway, the antennas worked, and we received beautiful signals.

Our second major difficulty was the elimination of the beat note when we connected in the General Radio Oscillator. Thanks to a radio Conference on the amateur rig with Vic Simas, we were able to track down and cure the difficulty. However, after about 36 hours of operation the Hewlett Packard frequency counter gave up the ghost, and again we were off the air! This time it took the combined efforts of Jim Crane and myself to get the frequency counter working . . . . The difficulty was finally traced to an open-circuited resistor in a plug-in "trigger unit." To compound our difficulties, there was no adequate description of the unit in the instruction room, we had no spare unit, and we couldn't get to the unit to check the components under operation because of inaccessibility. Jim Crane finally soldered in leads from the plug in socket and wired them to the plug in unit outside the counter. The failure was finally traced to an open resistor which was in parallel with a second resistor for adjustment to an exact value. The open could be only detected by lifting each resistor from the circuit and testing. Needless to say, it was a long slow search, requiring about 8 hours. We did get the repairs made, and the counter operating about an hour before passage. At about −5 minutes the overheat cutout in the counter kicked out, and only by holding it in with a screwdriver could we keep on frequency. At −1 minute we shifted to higher speed recording, and the pens stopped writing. After frantic shifting to several speeds, we finally got the pens to writing at maximum speed, so that the 5 minutes record used practically an entire roll of paper, but we did get a beautiful record. What an experience! Needless to say, we got everything ready by the next run, but it was a rough day.

In addition to trying to get the minitrack running, we had the communication team working hard on installing the communication equipment. Everybody really worked. I was proud of the whole group. As of this date, the low power transmitter (about 1 kw) is working, but there is trouble with a water coil in the high power rig. Both rhombics are up, and the teletype machines are clicking away like mad. The crew is trying to learn army communication procedure, which is something like learning the ancient Sanskrit in two weeks!

Similar activity was underway all along the Minitrack fence, with the result that long before the Soviet satellite ceased transmitting, five of the stations were capable of limited tracking at the Sputnik frequencies. The data they obtained, supplemented by data from amateur radio operators and visual and optical observers, although approximate, were sufficient to permit the computation experts to improve knowledge of the Russian orbit and to predict the course of the satellite days in advance. A measure of the spirited efficiency with which the Minitrack crews adapted their equipment to the Russian frequencies is found in the statement of Homer Newell, made during the opening session of the CSAGI conference in Washington, that it would take them "several months" to accomplish what in fact they succeeded in doing in a fortnight.[12] Amateur radio operators played a substantial role in this achievement. On Sputnik night the national IGY committee got in touch with the American Radio Relay League in West Hartford, Connecticut, calling on its 70,000 members—all "hams"—to lend assistance. Although ordinary radio sets were unable to acquire the signals, those equipped with short-wave receivers and beat-frequency oscillators—standard equipment among the hams—could do so. In a matter of hours amateurs in this country and abroad were picking up the signals and logging them. On Sputnik night they forwarded their findings to the National Academy of Sciences, where Porter, Berkner, Pickering and other scientists had established a temporary control room, setting it up with such speed that they were able to compute the orbit of Sputnik and inform the press of its whereabouts by 8 o'clock the next morning. Subsequently the hams communicated with the permanent control room at NRL.[13]

Word of the Soviet launch reached the headquarters of the Vanguard optical-tracking program at the Smithsonian Astrophysical Observatory in Cambridge at 6:15 Sputnik night. The Observatory Philharmonic Orchestra was holding its first rehearsal of the season, but one by one, as the session proceeded, members of the group quietly left the room. Whipple had been attending sessions of the CSAGI conference and was en route home from Washington. At the Cambridge Observatory his assistant, J. Allen Hynek, got the news in the form of a phone call from a Boston newspaper reporter, asking, "Do you have any comments on the Russian satellite?" An hour later Kittridge Hall, home of most of the tracking offices, was so ablaze with light that a woman living in the neighborhood reported that the building was on fire and a pumper and a hook-and-ladder went clanging to the scene.

No Baker-Nunn cameras were operational on Sputnik night, but with the aid of a hastily installed teletype machine and lavish use of the telephone, the SAO staff got the word to Moonwatch teams and astronomical observatories in this country and around the world. Fortunately some of the amateur units had undergone successful practice runs. During the night the Observatory received what Whipple described as "observations of a sort" and the early morning hours of Saturday brought "fairly good observations" from the Geophysical Institute in College, Alaska. From these data the Observatory was able to advise Moonwatch groups as to when and where they might be able to sight the satellite. The first confirmed observations came on 8 October. They were the work of Moonwatch teams in Sydney and Woomera, Australia. The first confirmed observation in the United States was the work of a team in New Haven, Connecticut. It came on 10 October. Thereafter observations came in steadily, with approximately 363 confirmed sightings, most of them by amateur teams, during the lifetime of *Sputnik I*. According to Whipple, none of these sightings was of the satellite itself. What his visual observers were seeing was the casing of the satellite's burnt-out carrier. This piece of hardware, the shell of the last stage of the launching vehicle, had gone into orbit with its payload and was chasing *Sputnik I* around the world. On the basis of reports pouring into the Cambridge center, Whipple concluded that the Russians had painted their payload black for reasons that were never made public.[14]

Although the first of the 12 Baker-Nunn cameras projected for the Project Vanguard optical-tracking stations had been completed some weeks before the Soviet launch, tests at the factory of its makers, Boller and Chivens in South Pasadena, California, had revealed defects, and the large and complex instrument was dismembered for repairs. On Sputnik night, consequently, the only Baker-Nunn in existence was "literally scattered all over the plant," [15] and some of its gears and other parts had been returned to contractors for refinishing or remachining. Even so, work on the camera was so far advanced that when on the night of 4 October news of the Soviet launch reached the people at Boller and Chivens, they hopefully started to assemble the camera for observation on the following night, only to desist after Fred Whipple informed them that at that time the Russian satellite could not be sighted from Pasadena. By the evening of 17 October the camera was in good operating condition and the orbit of the Russian satellite was within range of the California city. When the orbiting carrier-rocket of *Sputnik I* appeared, according to the Smithsonian Institution's report of the event,[16] "it looked like a large airplane light." So low was it orbiting that "one probably could have photographed it with a Brownie camera." The satellite carrier went from horizon to horizon in approximately a minute and a half. During this period, the Baker-Nunn picked up "four or five" pictures of it, and would have had more if the operators of the camera had been more experienced in the handling of their intricate instrument. During the next few days the press carried the first pictures ever made of an artificial moon in orbit around the earth. On the Thanksgiving day following, scientists associated with a Harvard-sponsored meteor project picked up pictures of *Sputnik I* itself, the actual payload, with two super-Schmidt cameras in New Mexico. Their achievement prompted Whipple and Hynek to institute an interim program at some of their optical-tracking stations and elsewhere, utilizing super-Schmidt cameras and cinetheodolites, along with two small missile telecameras, borrowed from Army Ordnance. Started during the lifetime of *Sputnik I,* this backup phototrack program would remain in effect until mid-1958, by which time the full Baker-Nunn network was in operation.[17]

The swift accumulation of limited but usable information from the world's first artificial satellite was not the only product of the struggle to track *Sputnik I.* For those who participated in the effort, it would prove to be an unforgettable experience in international camaraderie. The sleepless experts at NRL and SAO, the crewmen straining to convert their facilities to the Russian frequencies at the prime Minitrack stations, the ham radio operators glued to their earphones, the Army technicians manning the Mark II stations they had installed, the men and women making up Moonwatch teams around the world—in the years to come all these people would be proud to think of themselves as members of a rule-less and officer-less organization fondly spoken of as ROOSCH or Royal Order of Sputnik Chasers. The foremost practical outcome of their cooperative labors was that the

American tracking teams were ready when on 3 November 1957 the Russians sent their second satellite, *Sputnik II*, into orbit.

Unlike its predecessor, the second Soviet moon was not a special device, orbiting apart from its carrier. It was the last stage of the launching vehicle. Circling the world once every 103.7 minutes, *Sputnik II* had an apogee of 1,038 miles, a perigee of 140 miles. It remained in space 162 days, falling into the earth's atmosphere on 14 April 1958. Weighing at least 1,120 pounds, it carried the 11-pound test dog, Laika, in a sealed compartment, along with instrumentation for measuring cosmic rays, solar ultraviolet and x-radiation, temperature, and pressures. Although its transmitters functioned only seven days, they supplied the world scientific community with disclosures concerning the biomedical effect of space travel on animal life, solar influence on upper atmosphere densities, and the shape of the earth.[18]

For everybody connected with Project Vanguard the immediate post-Sputnik period was one of swiftly developing and often overlapping events. Some Americans had always been deeply interested in the earth satellite program: now the majority of them were. After 4 October 1957 the Vanguard scientists and engineers found themselves working quite literally in a goldfish bowl. Unofficially their mission ceased to be merely one of putting a payload in orbit during the IGY. It became instead an effort to salvage the national prestige. Their thinly funded, modestly conceived, no-priority undertaking had become the great white hope of a people profoundly wounded in its *amour-propre*.

Within hours after the first Soviet launch, the Senate Preparedness Subcommittee chairmanned by Lyndon B. Johnson initiated a "full, complete, and exhaustive inquiry into the state" of the nation's satellite and missile efforts. On 9 October Hagen and Admiral Bennett went "up the Hill" to tell the Vanguard story to attorney Edwin L. Weisl of New York, the Johnson subcommittee's chief investigator, and his staff. Accompanying them was Brigadier General Austin W. Betts of the Department of the Army, whose task was to answer questions concerning the possibility, then under intensive discussion, of using the Army's Jupiter C, a version of its intermediate-range ballistic missile, as the basis of a backup satellite-launching program for Project Vanguard. Most of the Senate investigators' questions reflected current criticisms of the manner in which the United States had handled its satellite program. Considerable discussion dealt with the President's order that the satellite effort be kept "separate and distinct" from the country's military missile effort. There were rocket men in and out of the Army who viewed this arrangement as an inadvisable "division of the indivisible." In answer to the Senate investigators' queries, Hagen and Bennett explained that "the decision" to separate the two programs arose from the fear that "the military program might be delayed if this were not done." They added that subsequent to the separate-but-highly-unequal decision, it had become "apparent that the Jupiter C missile of the Army" could be "used as a booster for an earth satellite. However, the time required to make the necessary modifications to the Jupiter C would not have resulted in a material saving in time and might have reduced the scientific value of the earth satellite." The investigators concluded the session with a request that the Vanguard managers supply them with a report on the background, status, and plans of the project. During the preceding summer, fortunately, Hagen had directed his aides to prepare a chronological history of the project. Within a reasonably short time, this and other pertinent material were on their way up the Hill, to be digested by the Johnson subcommittee staff in preparation for a projected series of hearings by the subcommittee itself.[19]

Eisenhower also requested a briefing, and a few days after *Sputnik I*, Hagen and William M. Holaday, recently appointed director of guided missiles for the Department of Defense, called at the White House for this purpose. The official record covers Hagen's subsequent briefing of the White House staff on 15 October, but it fails to fix the date of his earlier session with the President, and those involved no longer remember. Later events, however, place it on 9 October at the latest, possibly the day before, since on the 9th Hagen was tied up with the Senate investigators on the Hill. In a fifteen-minute presentation to Eisenhower, Hagen and Holaday stressed the then experimental status of the Vanguard program. TV–2, then being prepared for firing at Cape Canaveral, was not a complete Vanguard vehicle, consisting as it did merely of a Vanguard first stage and two dummies in lieu of the second and third stages. So far no complete Vanguard vehicle had been flight-tested, and the one scheduled for launching in December— TV–3—was still at the factory. Moreover, this first complete Vanguard was not a mission vehicle; it was a test vehicle, designed primarily not to orbit a payload but to measure the performance of the launching vehicle itself. Plans, however, called for TV–3 to carry a minimal payload, an instrumented 6.4-inch, 4-pound satellite. Preliminary calculations indicated that it could put such a satellite into orbit, but no guarantee to this effect was possible under the circumstances. In short, were TV–3 to accomplish its mission, the Vanguard people would regard their success, in Hagen's words, as "a bonus." [20] Having given the chief executive a realistic summary of the situation, Hagen and Holaday—and everybody else associated with the Vanguard program—were understandably startled and dismayed when on 9 October presidential press secretary James Hagerty informed reporters that during the forthcoming December, Project Vanguard would launch a satellite-bearing vehicle. "In May of 1957," the White House statement read in part, "those charged with the U.S. satellite program determined that small satellite spheres would be launched as test vehicles during 1957 to check the rocketry, instrumentation, and ground stations and that the first fully instrumented satellite vehicle would be launched in March of 1958. The first of these test vehicles is planned to be launched in December of this year."

It is worth noting that the White House news release was an accurate enough reflection of the Hagen-Holaday briefing of the President. It did not say that the Vanguard people were going to place a satellite in orbit in December; it said only that they "planned" to launch one of their satellite-bearing "test vehicles," a far less difficult procedure, especially for a group of men who had already put up two test vehicles in a row and were on the verge of improving this record by their successful launching of a third one, TV–2, on 23 October. In the emotionally overwrought atmosphere of the early post-Sputnik era, however, it was perhaps too much to expect reporters to distinguish between a promised launch and a promised orbit. Apparently few, if any, did. The press bristled with stories saying that before the end of the year America's answer to the U.S.S.R. satellite would be circling the globe. Hagen and his staff had no choice but to regard the ill-timed White House release, or more exactly the news media's interpretation of it, as a command; and all units of Project Vanguard braced for an accelerated effort beset with uncertainties.[21]

What must have been welcome news to many anxious Americans came five days after *Sputnik II* with an announcement from the Pentagon that the Army Ballistic Missile Agency (ABMA) at Redstone Arsenal in Huntsville, Alabama, a unit commanded by Major General John B. Medaris, had received permission to participate in the American satellite program on a backup basis. "The Secretary of Defense today," the department's 8 November release read in part, "directed the Department of the Army to proceed with launching an earth satellite using a modified Jupiter C. This program will supplement the Vanguard program . . . . The decision to proceed with the additional program was made to provide a second means of putting into orbit, as part of the IGY program, a satellite which will carry radio transmitters compatible with minitrack ground stations and scientific instruments selected by the National Academy of Sciences."[22]

To people close to the satellite program this announcement was no surprise. They had been expecting it since *Sputnik I*. Some of them had been discussing the feasibility of such a move since the fall of 1955 when the Stewart Committee rejected Project Orbiter, the Army's satellite-launching proposal, in favor of the Navy proposal that had become Project Vanguard. For Project Orbiter the Army-directed rocket team headed by Wernher von Braun had designed a four-stage launching vehicle, to consist of the liquid-fueled Redstone rocket, the Army's short-range tactical missile, and three

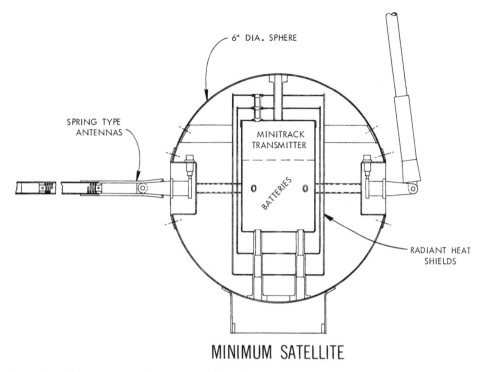

6" DIA. SPHERE

SPRING TYPE
ANTENNAS

MINITRACK
TRANSMITTER

BATTERIES

RADIANT HEAT
SHIELDS

MINIMUM SATELLITE

Sketch of the minimum six-inch satellite intended for TV–3.

solid stages made up first of clusters of Loki and later of scaled-down Sergeant rockets. When subsequently the Army rocket experts embarked on a series of tests designed to bring their nosecones safely back into the atmosphere during flight, common sense dictated that they use the four-stage vehicle they had planned for Project Orbiter as the basis for creating a suitable test missile. To this end they had developed what by 1957 was known as the Jupiter C, the "C" standing for "Composite Re-entry Test Vehicle." In this way the Army was able to carry on its vehicle development under military priority, an advantage denied the Vanguard program. Had the Jupiter missile been chosen in the first place as the IGY vehicle, it too might have had to undergo development outside military priority. Created by the Army in collaboration with the Jet Propulsion Laboratory of the California Institute of Technology, the Jupiter C was an elongated Redstone with three solid-fuel upper stages—two of them live, and the top one filled with sand to preserve the balance of the vehicle.[23]

As early as 1956 the performance of one of the predecessors of the Jupiter C convinced the experts at Redstone Arsenal that they already had a vehicle capable of putting a small satellite in orbit. Later reentry tests, carried on with the more elaborate Jupiter C itself, further strengthened their conviction. Late in 1956 the Department of Defense authorized ABMA to develop and fire twelve Jupiter Cs as part of the Army's nosecone re-entry development program. The first two shots, attempted in 1957, were failures, but a third, fired in August of that year, was such a definitive success that General Medaris, the ABMA commandant, ordered the reentry test program stopped and directed that the remaining Jupiter Cs—"nine precious missiles in various stages of fabrication"—be "held for other and more spectacular purposes."

By "other and more spectacular purposes," the dynamic ABMA chief meant a satellite launch. Twice, during the preceding year, he and his colleagues had requested permission to establish a backup satellite program; twice the Department of Defense had turned them down—but the Army missile team had no intention of taking "no" for an answer. Shortly after the successful reentry test in the summer of 1957, Medaris wrote Lieutenant General James M. Gavin, then Chief of Research and Development for the Army, that "we could hold two of the missiles in such condition that one satellite shot could be attempted on four months' notice, and a second one a month later." In his own vivid, engaging, partisan account of his stewardship of the Army missile program, Medaris confesses that in the summer of 1957 he was convinced that Project Vanguard's chance of effecting an orbit in the IGY was "so small as to constitute a ridiculous gamble." [24]

In October 1957 the Department of Defense was in the midst of a change of hierarchy. Secretary Wilson had announced his imminent resignation, and his designated successor, Neil H. McElroy, was making an "orientation" tour of the country's military installations preparatory to taking office. The fourth of October found McElroy at Redstone Arsenal. His party included Army Secretary Wilber M. Brucker, General Gavin, and other dignitaries. Their presence gave Medaris and von Braun an opportunity to renew their plea that the Army be given a role in the satellite effort. A briefing session and a tour of the arsenal were followed by an evening cocktail party. Hosts and guests were enjoying a relaxed chat when General Medaris' public relations officer hurried into the room. "General," he said, breaking into the conversation without apology, "it has just been announced over the radio that the Russians have put up a successful satellite! It's broadcasting signals on a common frequency, and at least one of our local 'hams' has been listening to it." There was a momentary silence. Then von Braun burst into speech. Medaris quotes the famous rocket scientist as exclaiming, "We knew they were going to do it. Vanguard will never make it. We have the hardware on the shelf. For God's sake turn us loose and let us do something. We can put up a satellite in sixty days, Mr. McElroy! Just give us a green light and sixty days." Von Braun talked on compulsively. It was some time before Medaris could interrupt long enough to observe that "sixty days" was too fast. To prepare vehicle and payload for launch, the Army and its working partner, JPL, would have to have "ninety days." Neither von Braun's sixty days nor Medaris' ninety could be called an excessively optimistic prediction. The Jupiter C was already a flight-tested vehicle, and there was no reason to believe that the modifications required to make it operational as a satellite-launcher would take much time. Moreover, it was a foregone conclusion that the Army team would have at its disposal the tracking system developed by the Naval Research Laboratory along with one or more of the scientific experiments prepared for Project Vanguard under the aegis of the National Academy of Sciences. Nothing could be more inaccurate than the subsequent popular impression that in a mere three months the Army team went through all of the time-consuming developmental work and testing necessary to produce an operational satellite-launcher.[25]

McElroy, of course, could make no commitments to the ABMA leaders, but by the time he and his entourage had departed on the following day, both Medaris and von Braun were under the impression that the "green light" would flash soon after the secretary-to-be took office on 9 October. Medaris' confidence took the form of immediate action. He ordered von Braun and his assistants to take "Missile 29," one of the Army's stored Jupiter Cs, off the shelf and start working on it. "I stuck my neck out," the ABMA chief would write later, but "I was convinced that we would have final word inside of a week, and that week was too valuable to be lost. If we *still* did not get permission to go, I would have to find some way to bury the relatively small amount of money we would have to spend in the meantime."

Even as Medaris and his aides pushed ahead in this informal manner, scientists connected with Project Vanguard were giving consideration to a cooperative Vanguard-Army venture that, in the opinion of some of them, just might permit the United States to launch a satellite within as little as thirty days. Basic to this scheme were calculations that Joe Siry of the Vanguard team had prepared. These indicated that the now tested and highly efficient solid-propellant third stage of the Vanguard vehicle could be fitted onto the basic Jupiter missile, thus providing a simple but powerful two-stage launching vehicle capable of establishing one of the small Vanguard satellites in orbit. In October Siry, Milton Rosen, and Commander Berg spent a day at Huntsville, where they presented Siry's figures to von Braun and his staff. The Army scientists agreed that Siry's scheme was feasible, and when the Vanguard men departed that night it was with the understanding that the two groups would pursue the matter further at a meeting of the Stewart Committee scheduled to be held in Washington a few days later. There, however, the matter ended. Siry's recollection, voiced ten years later, was that "after that first meeting it was somehow never convenient for us to get together again." Berg's guess, based on subsequent events, was "that apparently the scheme for marrying our rocket to theirs fell through because by the time we proposed it, the Army people were under the impression that they had a commitment from McElroy to go ahead with their Jupiter-C program. Under the circumstances a joint venture held no appeal to them, even though they conceded that Siry's calculations checked out." As for Rosen, his comment, also voiced a decade after the incident, was "that I don't really recall that occasion. Even if the Army had gone along with the idea, I don't believe I could have approved of it. No doubt the scheme looked good on paper, but difficulties involved in actually launching the proposed two-stage rocket would probably have put us behind rather than ahead of schedule. In those rather trying days, my feeling was that our best bet was to continue developing the Vanguard vehicle as planned and hope for the best."

One thing is certain. Throughout October the assumption that the Department of Defense was on the verge of authorizing an Army backup program prevailed at Huntsville. For General Medaris, to be sure, it was a less than comfortable period. Having ordered his men to go to work on the Jupiter C, he shortly found that his technically unallowed expenditures were becoming worrisomely large. When first one week and then three more weeks passed without a word from Washington, the general began awakening in the middle of the night and talking to himself. He would have rested better had he known that one of Eisenhower's last orders to Charles Wilson, his retiring defense chief, was a directive on the morning of 8 October "to have the Army prepare its Redstone at once as a backup for the Navy Vanguard." It was a month later, however, before the official directive establishing the Army satellite program, subsequently known as "Project Explorer," wended its way from DoD to Redstone Arsenal.[26]

It would be gratifying to report that the Vanguard people welcomed the Army to the satellite fold with expressions of delight and that the Army team refrained from gloating—gratifying but inexact. The Vanguard people detected in the situation causes for resentment that cannot be cavalierly dismissed as normal to interservice rivalry. Many of the difficulties confronting their project grew out of the government's decision in 1955 to keep the country's scientific satellite effort separate from its missile program. The Vanguard leaders can be excused for viewing the government's swift reversal of this position in the face of the Sputnik crisis as something short of an expression of faith in their endeavors. A further source of resentment lay in their conviction that the Army missile team had jumped the gun by preparing for a satellite shot years before getting authorization to do so. Apprised of this frequently advanced charge years later, former President

202

Eisenhower expressed surprise, saying "but that would have been a court martial offense!" As for the men in charge of the rocket team at Huntsville, Medaris concedes that their feelings were ambivalent. "We were angry and frustrated," he writes, "at having our country so badly outmaneuvered. On the other hand, we were jubilant over the prospect of at last being allowed to get our own satellite off the ground." [27]

During November the Johnson subcommittee investigators in Washington completed their preliminary study, and Room 318 of the old Senate Office Building became the scene of a series of open hearings and executive sessions dealing with the country's satellite and missile problems. Hagen's appearance before the subcommittee in late November had its light moments. At one point the noisy grinding of the television cameras brought complaints from some of the senators: they were having trouble hearing the soft-spoken Project Vanguard director. Chairman Johnson told the cameramen to do their work more quietly or get out. At another point a bulb suddenly tumbling from the chandelier gave rise to an exchange of pleasantries relative to "these strange flying objects" everybody seemed to be seeing in the heavens these days. One question came up repeatedly: could the United States have beaten the U.S.S.R. into space if the government had given Project Vanguard a higher priority? Quite likely yes, was Hagen's reply, provided, he took care to add, the project had been given also the things that go with a higher priority, namely "men, materials, funds." When Senator Estes Kefauver of Tennessee grumbled that "Well, Congress has given you all the money you asked for," Hagen patiently pointed out that this was one of those cases of things looking better than they were. He reminded the senator that for almost two years all of Project Vanguard's money had come out of the Defense Secretary's emergency fund. "The procedure," he explained, "was that we would go to the Department with a . . . request for funds for the remainder of that particular year, and some months, it would be 2 or 3 months after the request [before] the funds would be forthcoming." Congress' recent action in authorizing 34.2 million dollars to carry Vanguard to completion constituted the project's single contribution from that source, and even it was not a direct appropriation to the satellite project. Apparently the senator from Tennessee found food for thought in Hagen's words. At a subsequent hearing he saw to it that the subcommittee record included the statement published in a national magazine by Clifford C. Furnas, formerly assistant secretary of defense for research and development, that from the beginning Project Vanguard was "stymied by the chronic monetary constipation of the Armed Forces wherever expenditures which they consider non-military are concerned. Money was squeezed from odd corners of the various military budgets, and we got all the help which the National Science Foundation could give us. But the funds were dribbled out in such a manner that work was often slowed up for weeks and months at a time."[28]

While worried statesmen on Capitol Hill looked for answers to the country's space-age dilemma, the Vanguard Operating Group at Cape Canaveral strained to fulfill its commitment to put a payload in orbit before the end of 1957. For the members of the field crew the happiest outcome of the successful launching of TV–2 in late October was that it freed them to concentrate on preparing their first complete test vehicle, TV–3, for a flight test in December. The scheduled test of this vehicle, which would carry a small payload equipped with a beacon transmitter, would mark the first attempt to flight-test the second stage, the highly advanced liquid-propellant rocket that Martin's subcontractor, Aerojet General, had designed—and for many months the Vanguard vehicle experts had been struggling with a problem connected with the thrust chamber of the Aerojet rocket. To function successfully in flight the second-stage thrust chamber had to have a burning time in the neighborhood of 150 seconds, which is to say it had to be able to fire that long over and above whatever time it had been fired for testing purposes prior to flight. Static tests conducted at the Aerojet plant in 1956 and 1957 showed that in this connection a steel thrust chamber presented no difficulties. Indeed the one steel chamber fabricated at the factory eventually accumulated 600 seconds of firing time without any evidence of erosion. Unfortunately a steel chamber raised a weight problem beyond the capacity of the Aerojet engineers to solve in the limited time at their disposal. To get around this they fabricated a series of lightweight thrust chambers, using tubes of 5052 aluminum, hand-welded and wrapped with stainless steel wire. The aluminum chamber weighed twenty pounds less than the steel one, but during the summer of 1957 test firings yielded discouraging results. Four of the aluminum chambers developed internal leaks after 327, 240, 364, and 278 seconds of accumulated firing time respectively. Jim Bridger, the Vanguard vehicle chief, and his colleagues viewed a chamber lifetime of something less than 278 seconds as inadequate, and instituted an intensive search for a coating capable of extending the lifetime of the chamber, preferably to as much as 540 seconds. This proved to be a time-consuming process. Weeks of experimenting preceded the discovery that the application of a tungsten carbide coating to the chamber walls was the best available answer to the problem. In October the project managers ruled that beginning with TV–5 all second-stage thrust chambers must be so treated. Bridger would have liked to see a tungsten carbide-coated chamber in the second stage of TV–3, but that called for more time than Project Vanguard's accelerated post-Sputnik schedule permitted. For TV–3 the field crew used a second-stage thrust chamber that had been test-fired for only 50 seconds; hopefully it would last for another 150 seconds in

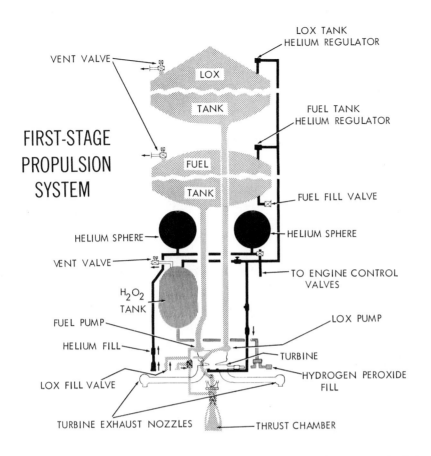

FIRST-STAGE
PROPULSION
SYSTEM

VENT VALVE

LOX TANK
HELIUM REGULATOR

LOX

TANK

FUEL TANK
HELIUM REGULATOR

FUEL

TANK

FUEL FILL VALVE

HELIUM SPHERE

HELIUM SPHERE

VENT VALVE

H_2O_2
TANK

TO ENGINE CONTROL
VALVES

FUEL PUMP

HELIUM FILL

LOX PUMP

TURBINE

HYDROGEN PEROXIDE
FILL

LOX FILL VALVE

TURBINE EXHAUST NOZZLES

THRUST CHAMBER

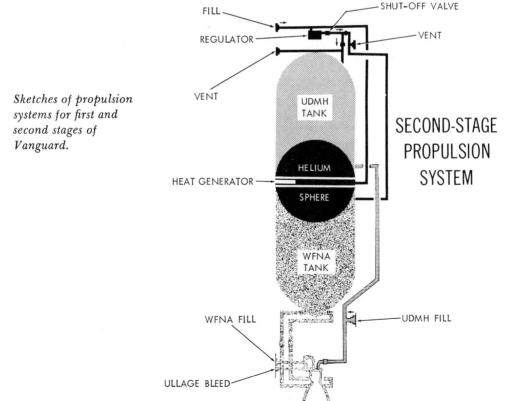

*Sketches of propulsion
systems for first and
second stages of
Vanguard.*

FILL

SHUT-OFF VALVE

REGULATOR

VENT

VENT

UDMH
TANK

SECOND-STAGE
PROPULSION
SYSTEM

HELIUM

HEAT GENERATOR

SPHERE

WFNA
TANK

WFNA FILL

UDMH FILL

ULLAGE BLEED

flight—assuming that the first complete Vanguard vehicle lifted off successfully.[29]

Compared with the difficulties the Vanguard field crew had experienced in preparing TV–2 for launch, those arising during the preparations of TV–3 were encouragingly minor. On 11 October the vehicle arrived at Hangar S, the project's permanent assembly building at the Air Force Missile Test Center. Inspection revealed a crack in the second-stage engine. This problem the Martin company took care of, promptly removing an engine from another test vehicle and shipping it to the field. Early in November TV–3 made the four-and-a-half-mile journey from Hangar S to launch complex 18A. Static firings of the two lower stages, erection of the vehicle on the launch stand, electrical system tests, instrumentation system tests, other preflight operations—all proceeded smoothly. The last weekend of the month found thousands of people converging on Cape Canaveral, ostensibly to witness America's first attempt to put an earth satellite in orbit, actually to watch the first test of the vehicle designed to carry out this job. Hopefully scheduled for Wednesday, 4 December, this test—as was true of all similar operations during this early period of rocket technology—should have been carried out in a quiet atmosphere. Unfortunately the Presidential news release of the previous 9 October [30] had been interpreted by the newspapers and then the public to mean that this was indeed Project Vanguard's first serious launch attempt. The fat was in the fire, and the field crew had no choice but to proceed as though this was indeed a scheduled attempt to launch a satellite. Before the weekend was over accommodations had become hard to find in the sparsely settled Cape Canaveral area, and local stores were out of binoculars. A *New York Times* reporter provided his readers with a colorful picture of the scene. "Last night from one of the coarse sandy beaches where the 'bird watchers' of the missile age watch the Cape Canaveral spectacles," he wrote on Sunday, the first day of December,

> the Vanguard tower was clear against a starry sky, two bright white lights glaring at its base and a red beacon shining at its top. This morning the alternating red and white horizontal stripes of the big crane stood out against the chill morning. From the beach, the Vanguard crane is one of a community of launching structures [service towers], some taller, some broader than others. But the Vanguard clearly has next billing at the sprawling missile theatre here . . . . The audience includes Project Vanguard officials and correspondents from as far as Europe.

For Florida the weather was disappointing. Sunday's temperatures were near freezing. Of greater concern to project officials were the wind velocities. These ranged from 20 to 30 mph. Field-manager Mazur and his staff could only hope for an improvement by Wednesday. According to NRL's aerodynamics specialists, a successful launching was unlikely in the presence of surface winds having a velocity in excess of 17 mph. Involved in these calculations was the "von Kármán effect," named for Theodore von Kármán, the scientist responsible for its formulation. The von Kármán effect refers to the tendency of air as it flows past an object and reaches the other side to curl or produce eddies. The eddies set up an oscillation that may be reinforced by the natural frequency of the structure. The effect of this on a rocket standing on a launch stand, especially after its service tower has been withdrawn some sixty minutes prior to flight, is a strong vibration with destructive possibilities. In an effort to minimize the von Kármán effect, Vanguard technicians equipped their seventy-two-foot vehicle with black rubber spoilers. These fin-like strips extended down the sides of the structure for about two thirds of its length. At the top of each was a protruding shoe, designed to catch downrushing air and strip off the spoilers at about a thousand feet altitude.[31]

Director Hagen was not among the project officials at the launch site. He remained in Washington, supervising activities in the Vanguard Control center at NRL. At the Cape his deputy, J. Paul Walsh, who was in charge of operations acted as project spokesman. On Monday he answered reporters' questions. One wished to know if the success or failure of the impending launch could be judged by whether or not the "baby moon" on the top of the vehicle went into orbit. "It depends on who's judging it," Walsh replied. "Don't misunderstand me. We'll be pleased if it goes into orbit. We'll not be despondent if it does not." The schedule called for the countdown to begin at 9 o'clock Tuesday evening, but satisfactory completion of some of the last-minute tests took more time than anticipated. It was 4:30 Wednesday morning, with giant searchlights bathing the rocket with a blue-white light, before the countdown began. At 10:30 that night, two holds and eighteen hours later, Mazur scrubbed the shot for three reasons: a frozen shutoff valve, fatigue on the part of his crew, and meteorological readings showing that winds in a jet stream located over the launching site had reached velocities considered marginal for flight firing.[32]

Wednesday night's cancellation of the initial attempt to launch TV–3 was followed by an announcement that the field crew would start another countdown late Thursday afternoon with liftoff scheduled for 8 a.m., Friday, 6 December. Speaking at a business meeting in Florida on Thursday, George S. Trimble, Jr., a Martin Company vice president, flatly asserted that the first complete Vanguard vehicle would not succeed in placing its payload in orbit. He based his prediction on "the prevailing mathematics of trial and error." According to these calculations three failures for every seven tries were normal in "this kind of testing experiment." At a news conference in Chicago on the same day the chairman of the IGY committee, Joseph Kaplan, was only a little more optimistic. Cautioning reporters about the "risk of failure in tomorrow's shot," he assured them that before the end of the International Geophysical Year on 31 December 1958, the United States "will have a full-fledged earth satellite in orbit." These last-minute efforts to prepare the American people for the worst are of interest in view of the events of the next twenty-four hours.[33]

The second countdown began shortly after 5 p.m. Thursday, approximately on schedule. Shortly thereafter a long hold became necessary because of delays encountered in verifying the operations of the vehicle controls system. Subsequent holds were of short duration and of no significance. By 10:30 Friday morning the countdown had reached $T-60$ minutes, the beginning of the final and critical phase of the procedure. At this point the big gantry crane began its slow withdrawal, leaving the vehicle standing alone on its flight-launch structure. A weather check, ten minutes later, showed winds of 16 mph at pad level with gusts up to 22 mph. For later Vanguard flight tests, the Martin Company would design a retracting launch stand that permitted the vehicle to lift off in surface winds up to 35 mph; but on 6 December 1957 the original stationary stand was in use and Martin studies had fixed the allowable ground wind for liftoff at only 17 mph. In higher winds the engine nozzle, as it rose from the clearance hole in the platform of the stationary stand, might crash against the surrounding piping. At $T-50$ minutes, in short, weather conditions were touch-and-go, but otherwise all looked well. At $T-45$ minutes the electronics telemetering crew in the backroom of the blockhouse began receiving "all clear" signals from the stations of the radio tracking network. Photographers in the employ of Pan American Airlines, responsible for range servicing and general engineering, were busily immortalizing the occasion, snapping pictures of equipment and individuals. At $T-30$ minutes fierce blasts from the bullfiddle warning horn on the launch pad sent people scurrying from the area. Some retreated to their assigned posts in the blockhouse, others made off in their cars to safely distant points. At $T-25$ minutes the heavy blockhouse doors clanged shut. The air of tension generated by the busy occupants of the building edged upwards from high in the direction of unbearable. At $T-19$ minutes the blockhouse lights went out, the "No Smoking" sign blinked on. A report that surface winds were now "fifteen knots" brought a shrug from Dan Mazur. The figure was high, but the trend was downward. Indications were that by liftoff, the wind velocities would be acceptable. At $T-5$ minutes, propulsion-expert Kurt Stehling detected a "quaver" in the voice of his assistant, Bill Escher, who was counting off the minutes over the public address system. Five minutes later Escher changed the count to seconds. At $T-45$ seconds, the so-called "umbilical cords" that supply the rocket right up to liftoff began dropping away. At $T-1$ second test conductor Gray gave the command to fire and Paul Karpiscak, a young Martin engineer, flipped the toggle switch on his oblique instrument panel. In the crowded blockhouse control room all eyes were on the big windows overlooking the pad. Sparks at the base of the rocket signaled that the pyrotecnic igniter inside the first stage had kindled the beginning of the oxygen and kerosene fumes. With a howl the engine started, brilliant white flames swiftly filling the nozzle and building up below it as the vehicle lifted off. The time was 11:44.559 a.m. Two seconds later, a scream escaped someone in the blockhouse control room: "Look out! Oh God, no!" To Kurt Stehling, his gaze on the spectacle outside, it seemed "as if the gates of Hell had opened up." With a series of rumbles audible for miles around, the vehicle, having risen about four feet into the air, suddenly sank. Falling against the firing structure, fuel tanks rupturing as it did so, the rocket toppled to the ground on the northeast or ocean side of the structure in a roaring, rolling, ball-shaped volcano of flame. In the control room someone shouted "Duck!" Nearly everybody did. Then the fire-control technician pulled the water deluge lever, loosing thousands of gallons of water onto the steaming wreckage outside, and everybody straightened up. The next voice to be heard in the room was that of Mazur, issuing orders: "O.K., clean up; let's get the next rocket ready." Already the stunned crew had taken in a startling fact. As TV–3 crashed into its bed of flame, the payload in its nosecone had leaped clear, landing apart from the rocket. The satellite's transmitters were still beeping, but the little sphere itself would turn out to be too damaged for reuse. It rests today in a file cabinet of the NASA Historical Archives, a battered reminder that "The best laid schemes o' mice an' men/ Gang aft a-gley." At the Vanguard assembly building, four and a half miles northwest of the blockhouse, Paul Walsh was on the phone to Hagen. The open hangar doors gave him a view of the launching pad. At $T-0$, he passed on the news: "Zero, fire, first ignition." His next statement was a single word: "Explosion!" At the Washington end of the line, project director Hagen was equally succinct. "Nuts!" he said.[34]

All components of the first stage of the Vanguard vehicle had functioned in a "superior" fashion during the successful launching of TV–2 in October. What had gone wrong with those same components during the flight firing of TV–3? Did the fault lie in the first-stage engine, the X–405 liquid-propellant engine developed by GLM's subcontractor, General Electric? Or did it lie in the other major component of the stage, the tankage built by GLM itself? During TV–3's two seconds of life after liftoff, its onboard telemetry worked. Consequently the General Electric and GLM investigators had on hand a collection of telemetered data concerning the behavior of the rocket that Walsh described as "worth its weight in gold." They also had ground instrumentation records and a series of photographic films of the disaster. Technicians of the two companies studied these, and came up with different answers. The Martin people traced what they called an "improper engine start" directly to a low fuel tank pressure which was responsible for a low fuel injector pressure prior to the start of the turbopump operation. The low injector pressure allowed some of the burning contents of the thrust chamber to enter the fuel system through the injector head. According to this version of the accident, fire started in the fuel injector before liftoff, resulting in destruction of the injector and complete loss of thrust immediately after liftoff. The General Electric investigators dissented. They traced the immediate cause of the explosion to a loose connection in a fuel line above the engine. Their reading of the telemetered and photographic data was that there was no "improper start." On the contrary, the engine had come to full thrust, only to lose thrust when a little leaked fuel on top of a helium vent valve blew down on the engine.

In a remote sense the General Electric investigators held the Martin work crew at fault. They claimed that members of the crew had used the fuel lines as "ladders" while working on the vehicle; hence the loose connection. At a conference attended by representatives of the companies and NRL, Milton Rosen, the project technical director, cut short what gave signs of becoming a heated argument. Conceding unofficially that the cause appeared to be "indeterminate," Rosen said the Project managers would accept GLM's findings. Although GE continued to hold to its position, its spokesmen appreciated the wisdom of Rosen's decision under the circumstances. In the aftermath of the TV–3 catastrophe, the time pressures on Project Vanguard were too severe to permit the luxury of a protracted family quarrel. In accordance with a specification change negotiated with Martin, GE increased the minimum allowable fuel tank pressure head of its engine thirty percent, and provided for manual override of the regulator to assure that this condition could be met. Time would confirm the practicality of this procedure. In fourteen subsequent flight and static firings of the first stage, the engine as altered started without incident.[35]

These technical matters were of no moment to the American people. A wave of outrage swept the country. "Failure to launch test satellite," the *New York Times* announced in big headlines, "assailed as blow to U. S. prestige." Senator Lyndon B. Johnson spoke for millions when he termed the situation "most humiliating." In New York City, members of the Soviet delegation to the United Nations asked American delegates if the United States would be interested in receiving aid under the U.S.S.R.'s program of technical assistance to backward nations. On the morning after the explosion sell-orders on Martin Company stock reached such proportions that at 11:50 a.m. the governors of the New York Stock Exchange suspended trading in it. When they permitted a resumption of trading at 1:23 p.m., the stock was at $36\frac{3}{8}$, off $1\frac{3}{8}$. At the end of trading it was $35\frac{1}{2}$, off $2\frac{1}{4}$. In the words

*TV–3 launch, 6 December 1957. Two seconds after launch, when
the vehicle was four feet off the pad, thrust ceased. TV–3 crumpled on
the pad and exploded.*

of Donald J. Markarian, the Martin Company's project engineer, "Following
the TV–3 explosion, Project Vanguard became the whipping boy for the
hurt pride of the American people." A few weeks after the event, Markarian
encountered trouble in getting a painter to do some work at his Baltimore
home. "Finally," the tall, dark, broad-shouldered engineer would recall later,
"one of the men I approached had the courtesy to level with me. 'To tell
you the truth, Mr. Markarian,' he said, 'I don't feel much like working for
anyone connected with Project Vanguard.' From the quantity of criticism
that came hurling at us, you'd have thought we had committed treason."
Here and there the voice of reason emerged. To Detlev Bronk, president of

the National Academy of Sciences, President Eisenhower put a pertinent question. "Were we Americans the first to discover penicillin?" he asked. "You know the answer to that, Mr. President," was Bronk's reply. "And did we kill ourselves because we didn't?" Eisenhower asked. Bronk allowed that the President knew the answer to that too. In a letter to Hagen, Vice President Nixon wrote that at "a time when you have been 'catching it' from all sides, I want you to know that I, for one, feel you should have every support . . . . Keep up the good work." Senator William F. Knowland of California pointed out that "Everyone understands we may have some failures in these launchings. The Soviet Union may well have had a dozen before they launched the first Sputnik." At a press conference in Washington, Hagen rebuked the newspapers for "excessive publicity." Murray Snyder, Assistant Secretary of Defense for Public Affairs, said there was no excuse for the exaggerated optimism with which reporters had covered events leading up to the unfortunate launch. "The Department of Defense," he said, "exercised great restraint in its announcements, stressing the fact that a preliminary test was involved and if the test satellite was put into an orbit that would exceed the purpose of the test." Reason, of course, had little popular appeal for the time being. What Americans wanted was an answer to the Russian Sputniks. With the failure of TV–3, Project Vanguard had ceased to be their great white hope. By the end of the year their attention was riveted on the efforts of the Army-JPL team to prepare Jupiter C for a satellite-launching attempt, tentatively scheduled for late January 1958.[36]

212

12

SUCCESS—AND AFTER

THE ADDITION to the American satellite effort of the Army team—the Army Ballistic Missile Agency (ABMA) at Redstone Arsenal in Huntsville, Alabama, and its partner, the Jet Propulsion Laboratory (JPL) of the California Institute of Technology in Pasadena—called for a series of high-level decisions in Washington. Some dealt with the scheduling of launches. This was an involved maneuver since both the Vanguard and Army teams would be using the same Cape Canaveral range. They would also be using much the same tracking, telemetry and orbit-computation systems, namely those that the Vanguard electronics experts had developed for their project, supplemented by microlock, a tracking and telemetry network that the Army had been using with its missiles since 1953. Because of these overlaps, sufficient time had to elapse between shots for AFMTC to prepare the requisite range support and for the units in charge of the electronics services to put their equipment in order. Complex as these arrangements were, most of them had been worked out by the end of 1957. By this time the Department of Defense had authorized the Army team to make two "earnest tries" to orbit a small cylinder-shaped satellite to be known as "Explorer," and the Naval Research Laboratory had transferred to the Army a scientific experiment that it had originally assembled for one of the Vanguard satellites.[1] Scientists at the Jet Propulsion Laboratory were modifying this instrumentation for use in the Army payload, and the Army's four-stage Jupiter-C missile had reached Cape Canaveral, where a field crew was readying it for erection on the firing table at launch complex 26A, one of the Redstone pads at AFMTC. In addition, the Army had selected 29 January 1958 for its initial launch attempt, with the understanding that the Vanguard team would try to put up another of its vehicles earlier that month.[2]

During the period covered by these develements, Dan Mazur and his field crew had pushed ahead with the Vanguard program. The TV–3 disaster on 6 December 1957 found a backup vehicle, TV–3BU, ready to leave the hangar. Its erection at launch complex 18A, however, had to await repairs to the Vanguard firing structure. Some of its components had been severely damaged by the explosion. By working around the clock, the crew completed the necessary repairs sooner than had been anticipated. Even so, most of December had passed before all three stages of TV–3BU could be placed on the firing stand and the long and complex prelaunch operations begin. On 23 January the first attempt to put up the vehicle failed when heavy rains shorted some of the ground instrumentation cables during launch countdown. The next three days saw three more countdowns, two of them almost completed—and three more scrubs. Finally, on 26 January, the Vanguard crew removed a damaged second-stage engine, ordered a replacement, and announced that it would make no further efforts to launch TV–3BU until 3 February.[3]

Since AFMTC could provide range support for only one shot at a time, this left the Army team with a discouragingly short period—less than a week—in which to make its first launch attempt. Fortunately its preflight preparations at Cape Canaveral were not excessively demanding. The Jupiter C had undergone several flight tests. Moreover, such static tests as the forthcoming attempt necessitated had been taken care of at Redstone Arsenal before the missile moved east. The major activities at the pad consisted of checking out the hazardous solid-propellant upper stages of the vehicle and of making sure that when the tub containing these rockets started to spin on top of the elongated Redstone booster, it would do so smoothly and without destructive vibration. Well in advance of the scheduled launch date, these procedures had been concluded, and preparations for the flight test itself were moving at a satisfactory rate.

Advance publicity was restrained and the launch date was withheld from the press until twenty-four hours prior to the anticipated firing. This policy reflected the determination of General Medaris, the ABMA commander, to protect the Army team as much as possible from the misleadingly optimistic type of attention that the press had heaped on Project Vanguard prior to the TV–3 explosion. Summoned to Washington in late 1957 and again in early 1958 to testify at the Johnson Senate subcommittee hearings on American missile and satellite programs, the general ducked the questions of reporters looking for more specific information. The Senate subcommittee itself gave him no problems on this score. When the matter of the Army's launch schedule came up, Cyrus Vance of the investigating staff informed Medaris that "I am not going to ask you about the date." Medaris' reply was "I am thankful for that, Sir." Appearing before the subcommittee on three occasions, the striking-looking ABMA chief was a colorful and articulate witness and both the senators and their staff handled him with a gentleness that must have made John Hagen, the beleagured Vanguard director, sigh with envy.[4]

On 29 January, launch day, the Explorer vehicle, its satellite and its field crew were ready, but disturbing reports were coming in from the AFMTC meteorologists. On the surface the weather was fine. Instrumented-balloon soundings, however, had revealed the presence high above the Cape of a jet stream, a swiftly-moving river of air, almost certain to destroy the missile. Heeding a teletyped advisory from his structural analysis engineers at Redstone Arsenal, Medaris decided to play it safe. Next morning's weather reading was slightly more encouraging. At noon he authorized the crew to begin an eight-hour countdown, only to call it off a few hours later following a report that the jet stream was again menacing.

At this point—Thursday evening, 30 January—time was running out for the Army team. Project Vanguard's next flight test of TV–3BU was still tentatively set for 3 February, and word from IGY headquarters in Washington was that the electronics units would need three days of preparation for it. The Army must either put up its vehicle on the following day—31 January—or hold off until the Vanguard team had completed its scheduled attempt. Medaris and his crew could only wait and hope. Next morning's 7 o'clock weather reading, as interpreted by the structural analysis engineers, was just favorable enough. "Things look good," it read. "The jet stream has moved off to the north, and by evening should be down to 100 knots." To Medaris that "still sounded like a lot of wind, but it meant the difference between a strain that we knew the missile could stand and one that was dangerous." In a now-or-never spirit, the ABMA commander set in motion another eight-hour countdown, prayerfully heading, as on the day before, for a firing at 10:30 that evening.

Beginning at 1:30 p.m., the countdown encountered no serious hitches. Late in the afternoon there was a half-hour hold to complete a number of operations that had fallen behind schedule, seemingly because crew members were still suffering from exhaustion after the exertions of the day before. Later they made up for the lost time. At 9:45 p.m., with the countdown exactly on schedule, there was a second hold when someone spotted a hydrogen-peroxide leakage in the tail of the missile. Workmen drained the line and stopped the leak. When at 10 p.m. the countdown resumed, it was only 15 minutes behind. At $T-12$ seconds—X–12, in Army terminology—the motors started to spin the top stages of the vehicle, technicians in the control room of the Redstone blockhouse transferred power from the ground power supplies to onboard sources, and at 10:48 p.m. the Jupiter C lifted off. It rose smoothly from its firing stand. A complex rocket, however, can fail even after a perfect start. There were jittery moments for the crew members while they awaited assurance that the upper stages had fired. For its later satellite-bearing missiles, ABMA would contrive an onboard system capable of igniting the upper stages automatically. No such system flew with the

Explorer I, *first United States satellite, was launched by ABMA on 31 January 1958 (above). The orbiting of the slim satellite was acclaimed by (left to right) Wernher von Braun of ABMA, James Van Allen of the State University of Iowa, and William Pickering of Jet Propulsion Laboratory.*

first Explorer missile because the ABMA scientists and engineers had not yet contrived a dependable one. Instead they had developed a method for ground-command firing the second stage at almost the precise second the missile reached its absolute apex following liftoff. This was done from the Redstone hangar. There, at an exactly and swiftly calculated moment, approximately 404 seconds after launch, a scientist pushed a button to fire the second stage. A simple timer then controlled the ignition of the third and fourth stages, operating so as to allow the full thrust of each to be applied before the next one fired.

Word that the upper stages had fired in response to ground command marked the start of still another period of nervous waiting and wondering. Was the satellite in orbit? Tracking stations on the West Coast would have to answer that. One or more of them would be the first to pick up the radio signal showing that the payload had circled the globe. General Medaris has described with understandable feeling the moment when "someone came up and shoved a piece of paper in my hands on which were these magic words: *Goldstone has the bird.*" This meant that at 12:51 a.m., 1 February 1958—one hour and fifty-three minutes after liftoff—a newly installed tracking station in California had picked up the satellite "on its first trip back around over the United States." The big headlines in that morning's newspapers invoked an all but audible sigh of relief across the country. The challenge of the Russian Sputniks had been met. America's first artificial satellite, *Explorer I,* was orbiting the earth.[5]

The Vanguard field crew's plans for making its next launch attempt on 3 February 1958, were a trifle optimistic. It was two days later before TV–3BU was ready to go, and again the Martin Company vice president's "prevailing mathematics of trial and error" (i.e., seven attempts to launch a satellite were likely to yield three failures) proved potentially valid.[6] The Vanguard team's fifth launch attempt turned out to be its second failure. This time, however, there was no spectacular explosion on the pad. The first-stage engine—the component involved in the TV–3 explosion—worked well. After a perfectly nominal start, the vehicle rose gently from its stand, but at about 1,500 feet altitude, after fifty-seven seconds of normal flight, a malfunction occurred in the control system. Subsequent investigation showed that spurious electrical signals had created motions of the first-stage engine in the pitch plane. These in turn developed dynamic structural loads, coupled with a rapid pitch-down that superimposed air loads of about the same magnitude. As a result, the vehicle broke up at the aft end of the second stage. It would appear that the "prevailing mathematics of trial and error" were no respecter of satellite-launching teams, for a month later the Army suffered its first failure. On 5 March the second Explorer missile lifted off well, but the fourth stage failed to ignite and the satellite, Explorer II, fell into the Atlantic. The Army team had now completed the two "earnest tries" originally authorized. Within hours after the failure of the second one, however, the Department of Defense dispatched orders for Medaris and his crew to prepare for flight a duplicate Jupiter C that ABMA had shipped to Cape Canaveral, just in case, and preparations for a third try had been inaugurated.[7]

It was now Project Vanguard's turn again. The vehicle this time was TV–4, identical with TV–3 and TV–3BU save for minor modifications that the manufacturers had made as the result of the lessons learned from the unsuccessful efforts to launch a satellite with the two earlier vehicles. The Army's failure to orbit its second Explorer was dispiriting to the members of the Vanguard field crew. If an old and tested rocket like Jupiter C could fail, they saw little reason to be sanguine about their relatively untried and far more sensitive and complicated bird. For TV–4 and all subsequent vehicles, the project managers instituted a change in the launching procedure. Instead of trying to run off the countdown on one day, they divided it into two phases, with the first one on T−1 day—the day before scheduled liftoff —and the second and longer phase on launch-day itself. Scientific considerations were a factor in the decision to introduce this procedure. Some of the experiments scheduled for future Vanguard satellites were unlikely to function effectively unless their carriers achieved orbit within specified hours of the day. The two-phase countdown, extending over two days, would make it easier for the crew to get the launch vehicle up within a time period limited by these considerations.

During the opening weeks of March, erratic weather and recurring mechanical and electronic problems aggravated a general, if rarely expressed, fear that TV–4 would go the way of its two immediate predecessors. Three canceled countdowns were the vehicle's record when on 16 March the crew embarked on the first phase of what was to be the final launch operation. This phase of the countdown moved to its conclusion without incident, and at four o'clock the next morning, St. Patrick's Day, the second phase began. At 6:50 a.m. there was a short hold: more electronic problems. At almost literally the last second, there was another and even shorter hold, or more exactly, a "stretch-out," when calculations showed that if the countdown concluded at that moment, *Explorer I* would be passing overhead just as TV–4 arched into the heavens. Passage of the Army satellite at that time, according to the electronics men, might interfere with the signals from the Vanguard payload. "An unprecedented event," Kurt Stehling would later call this moment: "I must confess that never in my earlier life did I expect to see the day when one would have to wait until satellite traffic in the sky was cleared for the launching of another orbiter." [8]

The seventeenth of March 1958, was a beautiful day. At 7:15:41 a.m.

after a nervewrackingly reluctant start that came close to carrying the launch stand itself into the air, TV-4 rose into a brilliantly sunny sky flecked with small white clouds. Now began the post-launch countdown. At the open-air communications center that the crew had improvised a thousand yards or so northwest of the blockhouse, Paul Walsh was again on the telephone to John Hagen in Washington. At approximately $T+1$ second he was shouting into the receiver, "There she goes, John . . . the flame is wonderful. Engine is burning smoothly." At $T+150$ seconds, he was telling the project director, "John, the second stage is separated." And at $T+490$ seconds, triumphantly, "John, the third stage has separated." There was reason now to believe that the payload was in orbit, but already long-deferred plans for victory celebrations remained in abeyance while, "like expectant fathers," everybody involved waited for confirmation from the Minitrack station at San Diego, California. In Washington, about 9:30 a.m., there was a clatter on the teletype linking the NRL control room with the California station. "We have got no signal yet," San Diego reported. Then: "Stand by, we may have it." The NRL operator tapped out a return message: "Give us the word ASAP [as soon as possible]." San Diego came back immediately: "This is it. We have 108.03 . . . also 108.00 [the two radio frequencies of the satellite] . . . . Good signal . . . no doubt . . . congratulations . . . ." In his cubicle of an office John Hagen put in a phone call to Alan Waterman, Director of the National Science Foundation. "It is in orbit," Hagan said. "You can inform the President." The little sphere that would be known as *Vanguard I* was circling the globe every 107.9 minutes—apogee, 2,466 miles; perigee, 404 miles; expected lifetime of satellite and its trailing third-stage casing, about 2,000 years. It goes without saying that, in the eyes of the public, the members of the Army team remained the heroes of the space age; it was they who had put up America's first satellite. But the Project Vanguard people had the satisfaction of knowing that in record time—only two years, six months, and eight days—they had developed from scratch a complete high-performance three-stage launching vehicle, a highly accurate worldwide satellite-tracking system, and an adequate launching facility and range instrumentation; more to the point, they had accomplished their mission, which was to put one satellite in orbit during the International Geophysical Year.[9]

By the time *Vanguard I* went into orbit, several changes had occurred in the administrative framework of the country's space effort, and an even more significant change was in the offing. The appearance of the Russian Sputniks in fall 1957 engendered a widespread clamor that the United States embark on a vastly expanded space program. Throughout the remainder of the year and into 1958, considerable discussion dealt with the question of who should operate this undertaking: one or more of the military

TV–4 was launched 17 March 1958, putting Vanguard I *into orbit.*

The 3½-pound, 6.4-inch satellite.

People on the ground sweat it out: top, in the blockhouse before launch; above, right, at the site watching the drama of the launch; above, at the site waiting to hear whether orbit had been achieved; right, Vanguard control room at NRL, showing the teleprinters for receiving data from the Minitrack stations.

services, some existing civilian agency such as the National Advisory Committee for Aeronautics or the Atomic Energy Commission, or a new organization, separate from existing governmental units? If a new organization, should its managers be military officers or civilians or both? Scientific opinion, emanating chiefly from the National Academy of Sciences and the American Rocket Society, favored a new agency under civilian aegis. A miscellany of bills, introduced in Congress during the first month of 1958, ran the gamut of possibilities. President Eisenhower proceeded slowly. Jolted by the intensity of public reaction to the Soviet space triumphs, he originated the office of Special Assistant to the President for Science and Technology, filling the position with James R. Killian, President of the Massachusetts Institute of Technology. Killian's duties were purely advisory, but the creation of his office foreshadowed the administrative changes to come.

While Congress and the Executive wrestled with the problem, Defense Secretary McElroy instituted first one and then a second reorganization of those elements of his department directly involved in the space effort. In November 1957, he named William Holaday, his special assistant for guided missiles, to the position of Director of Guided Missiles, with enlarged powers where both the missile and space programs were concerned. A few months later Congress passed a law authorizing the Defense Secretary to "engage in advanced research projects," and McElroy set up within the DoD a separate unit to be known as Advanced Research Projects Agency (ARPA). Holaday transferred his responsibilities to ARPA, and under the direction of Roy Johnson, a vice president of the General Electric Company, the new unit funded and supervised the country's space projects for a few months. That this arrangement was to be a temporary one became apparent soon after its inception. ARPA achieved formal status in early February 1958, but by that date the prevailing opinion in Congress and at the White House was that America's nonmilitary space program should be handled by a special civilian body set up outside the Department of Defense. In a message to Congress on 2 April, Eisenhower proposed establishment of the National Aeronautics and Space Agency with the proviso that the functions of the National Advisory Committee for Aeronautics be absorbed into this new agency. In July, Congress passed and the President signed the necessary legislation, and on 1 October 1958 the National Aeronautics and Space Administration began life, with Thomas Keith Glennan, president since 1947 of the Case Institute of Technology in Cleveland, as its first administrator.[10]

Under ARPA and later under NASA, the Army team continued to participate in the satellite effort throughout both the IGY and the one-year extension of it known as the International Geophysical Cooperation (IGC).

During this two-and-a-half-year period, the Army made nine attempts to launch a satellite, with four successes, an impressive percentage given the state of the art at the time. The Jupiter C put up *Explorers I, III,* and *IV.* A more sophisticated version of the Army missile, the Juno II, was the launching vehicle for *Explorer VII,* a 91.5-pound satellite established in orbit on 13 October 1959, about a month and a half before the conclusion of the IGY–IGC.[11]

Project Vanguard became a part of NASA on its inception. One of the NASA Administrator's first official acts, however, was to delegate management of the project back to the Naval Research Laboratory. In actuality, therefore, no significant administrative change took place, and the members of the Vanguard field crew continued to put up their vehicles in accordance with the one-a-month schedule established shortly before the launching of *Sputnik I.* The success of TV–4 in March left them with a spare vehicle on their hands since the Martin Company had assembled and the hangar crew had checked out a backup vehicle, TV–4BU, against the possibility of failure. In accordance with a suggestion from the IGY committee, TV–4BU went back to GLM so that technicians at the Maryland plant could remove some of the test instrumentation and convert it into an SLV—a production satellite-launching, or mission, vehicle—for use in a later flight. Field preparations for the next scheduled launching proceeded in an atmosphere of some tension. Although the vehicle involved, TV–5, was only a test vehicle, its mission was to try to orbit the first fully instrumented Vanguard satellite, a 20-inch, 21.5-pound sphere. By the first week of April the first stage of TV–5 was on the firing stand at launch complex 18A, but the pad managers postponed erection of the upper stages because of facts brought out in a motion picture of the TV–4 launch. The film revealed that at the liftoff of that vehicle, the hydraulic disconnects had not separated smoothly. The belief was general that the pull-away stand that the Martin Company was in the process of completing would take care of this potentially troublesome situation. Unfortunately the new movable firing structure would not be ready for some time. For the scheduled flight test of TV–5, the only course open to the crew was to make some modifications in the old stationary structure and hope for the best.[12]

Once again hope exceeded accomplishment. Launched at 9:53 p.m., 28 April 1958, the last of the Vanguard test vehicles lifted off without difficulty, but its intricately devised payload never reached orbit. Flight was normal through second-stage burnout. The second-stage sequence, however, did not complete itself electrically. Its failure to do so prevented arming of the coasting flight control system with the result that the third stage was unable to separate and fire. Three more failures followed. After a successful liftoff, the first mission vehicle, SLV–1, encountered trouble at second-

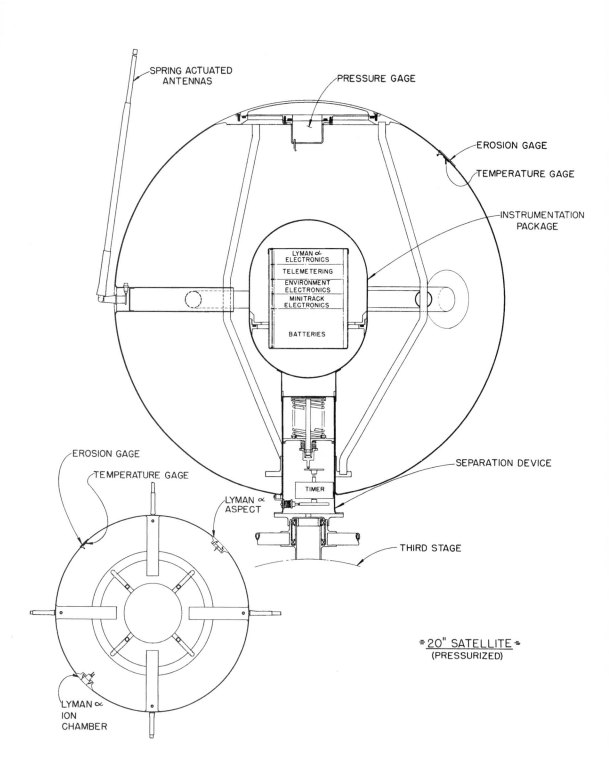

SPRING ACTUATED
ANTENNAS

PRESSURE GAGE

EROSION GAGE

TEMPERATURE GAGE

INSTRUMENTATION
PACKAGE

LYMAN α
ELECTRONICS

TELEMETERING

ENVIRONMENT
ELECTRONICS

MINITRACK
ELECTRONICS

BATTERIES

SEPARATION DEVICE

TIMER

LYMAN α
ASPECT

THIRD STAGE

EROSION GAGE

TEMPERATURE GAGE

LYMAN α
ION
CHAMBER

20" SATELLITE
(PRESSURIZED)

*Sketch and photograph of the first of the "operational" twenty-inch
Vanguard satellites. It was launched by TV–5, 28 April 1958, but failed to
achieve orbit because the third stage did not fire.*

stage burnout. At that point a disturbance in the control system had the
effect of launching the third stage at an angle of approximately sixty-three
degrees to the horizontal, thus precluding an orbit. SLV–2 also lifted off
satisfactorily, but the second-stage propulsion shut down after eight seconds
of burning. This action reduced the velocity of the vehicle to the point
where the third stage could not fire. Launched in September 1958, SLV–3
had the advantage of the new movable firing structure. During the liftoff
period, flight was normal or better than normal, but the performance of the

second stage was below the anticipated minimum. The burned-out third stage and the payload reached an altitude of nearly 265 miles, but the velocity was about 250 feet per second short of the 25,000 required to orbit.[13]

To the Martin Company men responsible for the reliability of the vehicle, none of these failures was a total loss. In every case they obtained sufficient telemetered and filmed data to spot what appeared to be the pertinent deficiencies and to make corrections. Indeed the care with which these follow-up procedures were carried out was one of the causes for the program's overall success. For the purpose of correcting deficiencies, the Martin design groups responsible for the various Vanguard subsystems kept in daily communication with their field counterparts by telephone and via direct teletype. In addition, they made constant use of a more formal channel for liaison and reliability follow-up—a form known as the "Discrepancy and Trouble Report," on which all malfunctions and actual or potential problem areas were recorded. At the plant the design men screened copies of these for problems requiring immediate action. Consisting of members from the engineering, manufacturing, quality, and procurement departments, a group called the Corrective Action Team met periodically to review each discrepancy report and to initiate corrective action or to verify action already taken as a result of the informal liaison maintained between shop and field. The Martin design groups learned as much, if not more, from success as they did from failure. One of the project's most successful flights, that of TV–4, for example, engendered more remedial action than any other single flight.

Thanks to Martin's intensive follow-up activity, Vanguard's fourth mission vehicle was considerably better than its predecessors. Launched on 17 February 1959, SLV–4 succeeded in establishing in orbit the 20-inch, 23.7-pound satellite now known as *Vanguard II*. Not that SLV–4 was a completely satisfactory vehicle in the eyes of its conscientious progenitors. Its payload, *Vanguard II,* exhibited an undesirable tumble rate. Telemetered data indicated that this had occurred because, following separation of the payload from the third stage, remnants of solid propellant remained in the rocket. When these ignited, they overtook and "nudged" the satellite, creating the undesirable tumble rate. Concluding that the trouble arose as the result of interference between the spring and a sharp shoulder on the separation device, GLM technicians placed in the separation hardware a thin metal sleeve. Their objective was to prevent binding in the succeeding mission vehicles, but although the identical problem did not present itself again SLV–5 and SLV–6 failed. SLV–5 was unable to orbit a 13-inch magnetometer and an expandable aluminum sphere because pitch-attitude control of the second stage was lost during first-stage separation. The resulting tumbling motion in the pitch plane aborted the flight. SLV–6 was

SLV–1 was the first of the "operational" Vanguard launch vehicles. Left, the second stage, the "brains" of the vehicle, being hoisted into place; below, the third stage being lowered onto the stack; bottom, the launch on 27 May 1958, which failed at second-stage burnout.

unable to orbit a 20-inch, 23.8-pound payload because a restriction in the propellant tank pressurant lines created a rapid decay of tank pressures immediately after second-stage ignition, followed by bursting of the pressurizing gas tank.[14]

The flight testing of SLV–6 on 22 June 1959 reduced Project Vanguard's arsenal to only two vehicles. One of these, TV–2BU—a left-over backup vehicle—was no longer usable. Set aside for exhibition purposes, TV–2BU stands today in rocket alley, between the Air and Space Museum and the Arts and Industries Building of the Smithsonian Institution in Washington, D.C. The other remaining vehicle was TV–4BU, the left-over backup test vehicle that the Martin Company had converted to mission status. The first two stages of this vehicle, as converted, reflected all the modifications that GLM had made in the Vanguard mission vehicles to correct deficiencies discovered in flight. In addition it was equipped with a new third stage. The top stage of all previous Vanguard vehicles had been the solid-propellant rocket motor designed and fabricated by the Grand Central Rocket Company. Grand Central's rocket consisted of a steel cylinder with a very thin—0.030-inch—skin, a hemispherical forward dome, and an aft dome fairing into a steel exit nozzle. At the center of the forward dome a shaft acted as the forward spin axis and supported the satellite. For TV–4BU the Allegany Ballistics Laboratory had built and tested a new solid-propellant third stage. Its shape was similar to the Grand Central rocket, but both its case and nozzle were made of glass-reinforced plastic. Theoretically, according to project engineers, as a satellite-launcher TV–4BU was as perfect as a vehicle of its thrust and configuration could be, and hopes were high at Cape Canaveral when in September 1959 preparations began for what was to be the Vanguard crew's final launch attempt. On the eighteenth of that month theory became reality. TV–4BU sent into orbit the fully instrumented 52.25-pound satellite now known as *Vanguard III*, along with its 42.3-pound third-stage motor case. The vehicle performed almost exactly as predicted. Thorough analysis of the flight brought forth no recommendations for change. In only fourteen launch attempts, the members of Project Vanguard had created an "operational" vehicle, capable of putting a 100-pound payload into orbit with a perigee of 180 miles.[15]

To appreciate more comprehensively the significance of this accomplishment calls for a glance at the work of the public information officers—the PIOs—assigned to the project. In charge of this activity throughout most of the program was Larry G. Hastings, who prior to joining Vanguard in the fall of 1957 had been with the Public Appearances Branch of the Office of the Secretary of the Air Force. A tall and generously proportioned man, Hastings' amiable manner and round face curtained a tough and agile

SLV–3 was launched 26 September 1958, from the new movable firing structure. The velocity of the third stage fell short of orbital velocity.

mentality. Assisting him was Mike Harloff, who came to Vanguard in May 1955, from the Headquarters of the Civil Air Patrol at Bolling Air Force Base.

Professionally both men were well seasoned, but a decade later both would admit in Hastings' words, that "Vanguard produced situations for which we could find no precedents in our experience." In the beginning, the two PIOs developed information procedures as they went along. "A project official might give us a ring," Harloff has recalled. " 'Fellows,' he'd say, 'here's a new problem. How do we handle it?' We had to come up with an answer, so we'd say 'Handle it this way' or 'Handle it that way.' Then and there the 'this way' or 'that way' became public relations policy."

Many problems were bound to arise from the schizophrenic nature of the program. On the one hand Vanguard was a part of the International Geophysical Year. As such its operations and its scientific findings were, so to speak, public property. On the other hand national security required that some elements of the undertaking, notably the components of the launching vehicle, be withheld from unauthorized scrutiny. In their effort to preserve the fragile line between what could be told the public and what could not, the information officers had the able guidance of James J. Bagley, head of the Security Review Branch of NRL and his assistant, H. W. "Ott" Ottenstroer. For each launch at Cape Canaveral, the Vanguard Project set up a crude communications control center for Project officers and the PIOs in and alongside a wooden shack—"of outhouse-dimensions," according to Harloff, and some 1,200 feet from the Vanguard pad, well within the danger zone. Here they took turns manning the phone over which they relayed information for dissemination to reporters covering the event both at the Cape and at the NRL news center in Washington. On these occasions, patient and understanding Jim Bagley was on hand. Every now and then the NRL security expert would tap the shoulder of the PIO at the phone, a signal that the information he might be about to relay for use by the reporters might trespass on classified domain.

Even had Project Vanguard been free of security elements, its scientific status would have been a source of friction to the news media at times. Since Vanguard was a research and development program, its hoped-for accomplishments could not be reliably forecast. "To ask us when we are going to put up our next satellite," director Hagen once remarked, "is at this point somewhat like asking medical researchers when they are going to find a cure for cancer." Like all experimenters, the Vanguard people could not say for certain what they were going to do until they did it. Obviously the policy of refusing to make public their unstable launch dates in advance was amply justified in those early days of the space age. It was scarcely calculated, however, to make life easy for Hastings and Harloff, one of whose

230

Vanguard II *was put into orbit by SLV–4 on 17 February 1959. Shown are the satellite in position, the launch, and "birdwatchers": left, Robert Schlechter, head of the Martin Co. field crew; center, Captain Peter Horn, Director of NRL; and, far right, Richard Porter of GE and Daniel Mazur.*

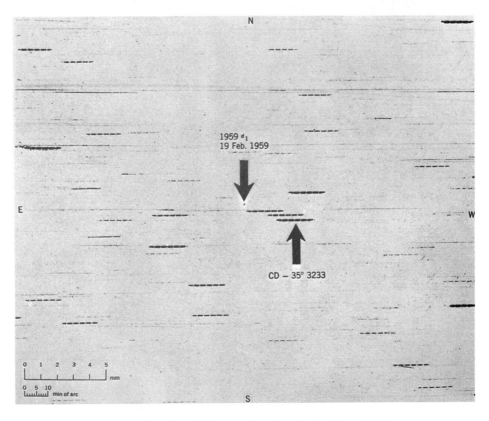

This photo of Vanguard II *in orbit was taken by the Smithsonian's Baker-Nunn camera emplaced at the combined radio and optical tracking station at Woomera, Australia.*

jobs was to cope with the gripes of some reporters, plagued by deadlines and editors hungry for "hard news" about the just-dawned space age.

A painful example of the tensions inherent in this situation is found in one of the misunderstandings that arose following the explosion of TV–3 in December 1957. At that time, one reporter seems to have made substantial effort to set the record straight. Writing in England's *Manchester Guardian*,[16] Alistair Cooke pointed out that true to their established practice, the Vanguard people had NOT announced the "great event" in advance. The premature release of the scheduled launch date of TV–3 was the result of a "leak," and NOT an announcement. Lyndon Johnson, the then Democratic leader of the Senate, clearly reflected worldwide feelings when he was quoted by Cooke as saying, "I shrink a little inside of me when the United States announces a great event and it blows up in our face. Why don't they perfect the satellite and announce it after it is in the sky?"

Launch of Vanguard III on 18 September 1959 by TV–4BU, ending Project Vanguard's flight program. Below, sketches of the three successful Vanguard satellites.

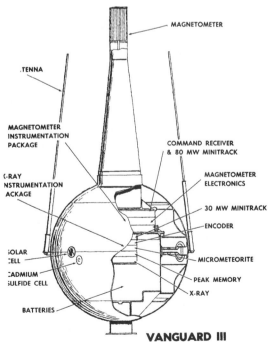

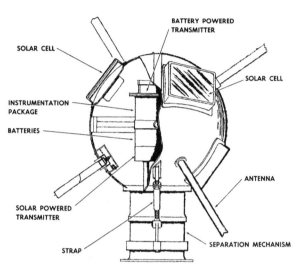

SOLAR CELL

BATTERY POWERED TRANSMITTER

SOLAR CELL

INSTRUMENTATION PACKAGE

BATTERIES

ANTENNA

SOLAR POWERED TRANSMITTER

STRAP

SEPARATION MECHANISM

VANGUARD I

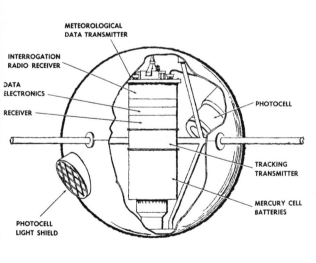

METEOROLOGICAL DATA TRANSMITTER

INTERROGATION RADIO RECEIVER

DATA ELECTRONICS

RECEIVER

PHOTOCELL

TRACKING TRANSMITTER

MERCURY CELL BATTERIES

PHOTOCELL LIGHT SHIELD

VANGUARD II

MAGNETOMETER

ANTENNA

MAGNETOMETER INSTRUMENTATION PACKAGE

X-RAY INSTRUMENTATION PACKAGE

COMMAND RECEIVER & 80 MW MINITRACK

MAGNETOMETER ELECTRONICS

30 MW MINITRACK

ENCODER

SOLAR CELL

CADMIUM SULFIDE CELL

MICROMETEORITE

PEAK MEMORY

X-RAY

BATTERIES

VANGUARD III

A source of unending irritation to the Vanguard team, and the PIOs in particular was the practice of some scientists not connected with the project to talk freely when approached for comment on Vanguard. Since they seldom, if ever, had access to the full picture of what was going on, their remarks were often misleading and sometimes unbelievably bizarre. To counter this stream of incorrect speculation, the information officers evolved techniques designed to eliminate error and to discourage speculation by placing correct, unclassified information in the hands of reporters as rapidly as circumstances permitted. To this end they prepared a simple, yet specific information plan for each launch for use by NRL and DoD personnel. This contained instructions as to what project official should be phoned for what type of data. In addition, they prepared in advance for issuance to newsmen a series of so-called "contingency statements." One such statement, for example, began with the words, "The Vanguard rocket was launched at _____ today. _____ (seconds) (minutes) after launch it (exploded) (fell back) (achieved orbit)," etc. Another statement was designed to take care of delays in meeting a scheduled launch date, this statement containing a blank space in which to record the reasons. After a launch or a postponement, when a reporter called in, he was quickly provided accurate information, based upon the appropriate contingency statement.

After the explosion of TV–3, the job of rebuilding the shattered public image of Project Vanguard was one long uphill climb. The orbiting of *Vanguard I* on 17 March 1958, of *Vanguard II* on 17 February 1959, and of *Vanguard III* on the following 18 September—even together, these successes failed to blot from the public's consciousness the picture of TV–3 bursting into raging flame on its launch pad. On the eve of TV–4BU, the launch that successfully resulted in *Vanguard III*, one reporter filed a story, the lead of which opened with these words: "Another ill-fated Vanguard stands poised on its pad at Cape Canaveral." Chancing to meet the author of this piece shortly after its publication, Hastings lost his customary geniality and heatedly said, "You must have knowledge that none of us on the Project have, and since you seem to have this advance information that the launch will fail, I think you have a duty as a citizen and taxpayer to call Washington and advise them that this launch should be scrubbed." So taken aback was the reporter that his only reaction was a feeble, "Say, you're mad about this, aren't you?"

During the life span of the project, the public relations officers undoubtedly had much to be "mad" about. "Our main problem," Hastings has commented, "was the fact that Vanguard had the unique position in those early days of the space age of being a public, or basically unclassified project. Vanguard was the only 'open project,' so it bore the brunt of the national displeasure with early space failures." [17]

Under these circumstances, the Vanguard public information officers could only do their best, from day to day, to tell the true story as it developed, taking what satisfaction they could from the knowledge that just as the Vanguard scientists and engineers were pioneering space-age hardware and procedures, they were also among the pioneers in space-age public relations.

Following the launching of TV–4BU in the fall of 1959, America's first purely scientific satellite program came to an end. In the language of officialdom it was "phased out" with practically all the NRL members of the project assuming positions with the various research and development programs of NASA. In the minds of those who were with the project more or less from start to finish, it remains vividly alive to this day. Its annual dinners on the seventeenth of March, anniversary of the launching of *Vanguard I*, draw in the neighborhood of a hundred of the team members, as well as veteran reporters who had sympathized with their efforts, to swap cherished reminiscences and to replay old Vanguard jokes. Dan Mazur speaks for all of them when he says that "for the great majority of us old Vanguard hands, putting up those rockets was never a job. It was a way of life." [18]

13

THE NATIONAL ACADEMY OF SCIENCES AND THE SCIENTIFIC HARVEST, 1957-1959

FOR THE NATIONAL Committee on the IGY and the Academy's satellite panel, the United States government's determination in fall 1957 to accelerate the American launching program had caused simultaneously satisfaction and apprehensiveness. Glad as scientists were to have their cherished project given a priority long denied it, they were uneasy lest haste to "get something up there" result in shortcuts that would lessen the quality of the scientific returns.[1] The intensive work put into tracking the two Sputniks and extrapolating scientific data from them proved the reliability of the Minitrack system, despite the differences in the radio frequency the Russian transmitters employed. It also established the utility of Moonwatch and Moonbeam operations and within a few weeks netted some useful geodetic information.[2] But the mandate to hurry up American launchings placed an additional burden on the scientists responsible for other phases of the American program.

In late October 1957 while waiting for official word of what the White House would authorize as a backup for Project Vanguard, the Working Group on Internal Instrumentation discussed the possibilities for the future. Repeating the suggestion that the panel's ad hoc group had tendered to the USNC nine months before,[3] Homer Newell opined that the United States should establish an "Astronautical Laboratory" under civilian control to direct a long-range space program. Here was the second mention of the scheme that would become reality a year later with the organization of the National Aeronautics and Space Administration. Meanwhile men at the Academy had to struggle with the problems immediately confronting them. Anticipating DoD consent to the use of the Army's Jupiter-C rocket as a satellite launcher, the panel had to decide which experiments to switch to the Explorer and what redistribution to make of the remainder among Vanguard satellites. With a bigger payload in prospect, there was also the question of whether to sponsor the development of further experiments for installation in IGY birds. And should the panel support ionospheric studies that would require transmitters and receivers operating on longer wavelengths than the 108 mc chosen for Vanguard? At the same time the IGY secretariat must issue reports to other National Committees on orbital data acquired from Sputnik and forward for CSAGI approval Fred Whipple's proposal that the nomenclature of IGY satellites consist of the year, a Greek letter, and, when appropriate, a number indicating the degree of brightness for each in turn—for example, 1957 Alpha for the first Sputnik. The strong probability that Soviet scientists would not release their findings until they had completed a thorough analysis led the panel and USNC, on the other hand, to postpone overtures for an exchange of technical and scientific data with the U.S.S.R.[4]

Formal authorization of two Explorer shots as a backup for Vanguard came in early November. Pickering had pointed out to the satellite panel six months before that the State University of Iowa experiment would lend itself fairly easily to use in the Jupiter-C payload,[5] but now he warned that the rapid spin rate of the Explorer satellite would jeopardize the performance of George Ludwig's cleverly designed tape recorder and playback mechanism. While the Jet Propulsion Laboratory stood ready to make the necessary adaptations, the time available would be too short to overcome that hazard in an Army satellite if it was to be launched in January. The only way to proceed was to omit the command receiver and the memory device and rely instead upon continuously operating telemetry to relay signals to ground stations. Sacrifice of the storage device would mean the loss of much of the scientific data, since ground receivers could record them only when the satellite was passing over the tracking stations. This circumstance might defeat one of the primary purposes of the experiment, namely the determination of the latitude effect of cosmic radiation. Two factors, however, tipped the balance. Of the packages planned for the first four Vanguard flights, only two were in the last stages of testing at the Naval Research Laboratory, one containing the Lyman-alpha and environmental studies, the other the cosmic ray observations and meteoritic measurements; the former would need more extensive and therefore more time-consuming changes to fit into the Explorer configuration than would the latter. The second inducement was the Army's promise to supply at the earliest possible moment a Jupiter-C vehicle so modified as to accommodate the data storage equipment omitted from the first Explorer. As the Army satellite was to remain attached to the casing of the fourth-stage rocket after burnout, the eighty- by six-inch cylinder-shaped body presumably would be no more difficult to track optically than the twenty-inch spherical Vanguard. So the TPESP voted for the transfer, provided Van Allen and his colleagues concurred. Van Allen was on an icebreaker in the South Pacific. The Navy was unable to communicate with the ship by radio, whereupon someone suggested sending him a Western Union telegram. To everyone's astonished amusement, the wire, relayed via Australia, reached him and he immediately sent back his approval of the switch. Part of the instrumentation for measurements of interplanetary matter was also to go into the first Explorer.[6]

The press version of the transfer caused some heart-burning at the Naval Research Laboratory, for when Robert Baumann, who had done much of the work on the engineering layout of the instrument package, handed the pot over to William Pickering of JPL, the newspaper story gave no hint that the Vanguard team had spent months in working out reliable temperature controls, in testing the scientific instrumentation, and in producing the ingeniously miniaturized package. The caption of the accompanying photograph labeled the pot in Pickering's hands: "The 20-pound satellite the Army hopes to launch."[7] So far from realizing that Vanguard had contributed anything at all to the later success of Explorer, much of the American public thereafter assumed that the Army, notably Wernher von Braun with a minor assist from JPL, had prepared the Explorer payload from start to finish and in less than eleven weeks. Inasmuch as George Ludwig and Jet Propulsion Laboratory engineers undertook the redesign and the testing, the assumption was unjust to them also.

The panel meanwhile had little choice about reshuffling the packages to fly in the Vanguard satellites. In keeping with earlier plans, the gauges for environmental studies and the ionization chamber for the Lyman-alpha experiment should have the first priority. At Newell's suggestion, however, with Herbert Friedman's endorsement, the panel approved preparation of a solar x-ray experiment as an alternative to the Lyman-alpha, since the substitution would necessitate only minor changes in the ionization chamber and could easily constitute package Ia; in fact, extreme solar activity at the time of launching might make Ia more valuable than package I. The cloud-cover experiment was to go into package II: the magnetometer and the NACA inflatable sphere, provided they were ready in time, were to be the next flown. Suomi's radiation balance equipment was to make up package IV. If, before the expiration of the IGY, additional launchings were to provide space for other projects, one or more of those on the backup lists—Singer's, JPL's, or the heavy nuclei experiment under development by Martin A. Pomerantz of the Bartol Foundation and Groetzinger of RIAS—would be the logical choice. Just as the panel was unwilling to recommend cutting short the tests of the scientific instruments to be flown, so, with four of the eighteen months of the IGY already gone, members saw the impracticality of looking for new, still undeveloped, onboard experiments. They stood by that decision even after *Sputnik II* carrying the dog Laika on her week-long journey evoked questions about expanding the American program to include experiments in the life sciences. Although a biologist at the National Institutes of Health had already submitted a proposal to study the effects of radiation on yeast cells in the vacuum of space, and although the Vanguard scientific group thought the package would fit into the 6.4-inch satellite, the panel concluded that that experiment, like other more complex new schemes, would have to await a post-IGY program.[8]

Nevertheless, as the Soviet satellites with their 20- and 40-mc radio transmissions opened up a unique opportunity for radio propagation studies and ionospheric research, the TPESP believed it could not ignore that promising field of investigation. Accordingly it set up a Working Group on Satellite Ionospheric Measurements under the chairmanship of Alan Shapley of the Bureau of Standards to recommend particular projects. Most of the studies would not require changes in the instrumentation within the satellites and would rely on the use of ground receivers attuned to the longer wavelengths of Russian radio transmissions. Although the panel envisaged the possibility of later putting into American satellites 20- and 40-mc transmitters in addition to the far more accurate 108-mc—an eventuality that materialized in October 1959 in the last IGY bird put into orbit—the bulk of the work would concentrate on the data to be acquired from Sputniks. The ionospheric studies were thus ancillary to, rather than an intrinsic part of, the American satellite program. Still, in early 1958 the panel endorsed eight ionospheric projects and undertook to obtain grants to support them. Added to other expenses unforeseen earlier, the cost increased the demands on the IGY budget.[9]

Money indeed posed a problem to the United States National Committee at every turn in the months following the appearance of *Sputnik I*. Within the first three weeks, Newell told the panel that the costs of tracking and analyzing the telemetry signals from Sputnik were outrunning Vanguard's financial resources, while Whipple estimated that Moonwatch would have to have at least $50,000 more a year, and as much as $200,000 might be necessary to expedite delivery of the Baker-Nunn cameras. The IGY committee must also find money to cover the expense of completing the engineering and testing of instrumentation for backup experiments. Changing over the cosmic ray apparatus to fit Jupiter C would alone cost about $161,-200, considerably more than the $106,375 allotted to construction of the original. There was also the matter of solar cells. Although the Explorer would have to depend on conventional batteries inasmuch as redesign of the circuitry in its payload would delay an early flight, the panel was gratified to learn that the six-inch grapefruit to be launched by the first complete Vanguard test vehicle was to carry six solar cells. The wisdom of providing for solar power in future American satellites seemed self-evident, despite the additional cost consequent upon the longer period of time during which radio tracking stations and data reduction centers would have to operate.[10] New expenses meanwhile were growing out of the necessity of mailing out from the Academy thousands of pieces of literature and individual letters in answer to inquiries from people all over the country; good public relations forbade ignoring either the school child's or the influential citizen's request for information.

On top of all these demands, in the opinion of the Academy's IGY Committee, larger sums should go into final interpretation of the scientific findings after reduction of the raw data. John A. Simpson, professor of physics at the University of Chicago, called this last phase all-important, as it concerned "the truly scientific aspects of the work." Foreign scientists, Simpson contended, could exploit the data assembled by American experimenters unless generous grants enabled the latter to spend time on "the fundamentals of research growing out of the IGY program . . . . Many new scientific discoveries await the full analyses of these data." [11] The United States must not neglect the final harvest. Hence, whereas the TPESP concluded that a $2.2 million supplementary appropriation would suffice, the National Committee believed $3.2 million necessary. Alan Waterman, however, remembered the assurances he had given Congress in 1956 that the universities would meet most of the costs of interpreting and publishing scientific results; the Science Foundation submitted a request for $2.2 million in January 1958. On 31 March the President signed the bill appropriating an additional $2 million for the IGY, and somewhat later the USNC itself pared $294,334 from the amount allotted to the expanded and accelerated program. By then the United States had three satellites orbiting the earth.[12]

Rejoicing over the triumphant flight of *Explorer I* on 31 January 1958, the much less touted orbiting of *Vanguard I* forty-five days later, and *Explorer III*'s performance on 26 March calmed public furor over the American program without noticeably lessening popular interest in what would come next. If much of the public was primarily eager to read of or see on TV further American exploits, and if Moonwatchers and Moonbeamers waited impatiently for chances to exercise their tracking skills on new satellites, scientists were above all anxious to learn what the accumulating data relayed to earth added up to. By 3 April, when the Academy's panel met to assess accomplishments to date and consider future plans, few precise data were on hand. Those from the cosmic ray apparatus in *Explorer I*, 1958 Alpha, were confusing and, because it had not carried a storage device as originally planned for Vanguard, about eighty-five percent of the signals were lost, not received at ground stations. Most of the information obtainable from the scantily instrumented six-inch Vanguard, 1958 Beta 2, had to derive from analysis of its orbit and was still incomplete. Explorer II had failed to orbit. Although the instruments in *Explorer III* were working fairly well and its tape recorder and storage mechanism were enabling Minitrack and microlock stations by interrogation to receive about eighty percent of the telemetered signals, the results were as puzzling as and largely duplicated those from 1958 Alpha. Tentative reports on Dubin's experiment were interesting but still inconclusive. Meaningful interpretations of what the American satellites were revealing would have to await more intensive study.[13]

Dubin, however, was able to present at the Second Astronautics Conference in April a paper entitled "Cosmic Debris of Interplanetary Space" in which he discussed his initial findings. The signals recording the quantity, spatial distribution, and size of interplanetary matter colliding with *Explorer I* and *Explorer III* were easier to understand than those coming from the cosmic ray apparatus. The gauges for micrometeoritic measurements indicated fairly clearly that the average influx of particles of as much as ten microns in diameter was not more than one per thousand per square meter per second, while the influx of particles with diameters of four to nine microns was about ten times greater. Readings from twelve days of operation of the equipment in *Explorer I* showed very much higher rates of impact during about eight hours of every twenty-four than occurred during the remaining sixteen hours. The experimenters' hypothesis ran that "meteor showers" accounted for the difference, an explanation substantiated by ionospheric observations made at several ground stations, notably at White Sands, New Mexico. Extrapolation then permitted estimates that the earth may have a daily accretion of up to 10 million kilograms of cosmic dust. Later studies would confirm this thesis.[14]

Van Allen and his associates, on the other hand, were excited and baffled by the data coming in from the 1958 Alpha and Gamma satellites. At altitudes below 1,000 kilometers, the readings obtained from the Geiger counters were consistent with known theories of cosmic ray activities, but above 1,000 to 1,200 kilometers very high counting rates occurred and then at periods fell abruptly to essentially zero. At a special session of the American Physical Society at the Academy on 1 May Van Allen gave a paper in which he described the enigma. The cause, he noted, might be malfunctioning equipment, but that explanation seemed invalid because the instrumentation in *Explorer I* differed from that in *Explorer III;* the latter carried a tape recorder. Possibly the satellites passed through regions to which few cosmic rays could reach, but Van Allen thought that "extremely unlikely." The only remaining explanation, and the one Van Allen concluded must be correct, was that the Geiger counter tubes encountered such intense radiation during the high-altitude portions of the orbits that the detectors had to operate above the overload level, greater than 35,000 counts per second. Analysis indicated the existence of radiation consisting in part of energetic particles, presumably protons and electrons, in geomagnetically trapped orbits. Further exploration of this phenomenon would greatly help scientists to understand it fully.[15]

Explorer IV, launched in late July at a fifty-degree inclination to the equator in order to establish a different orbit, consequently carried one Geiger counter shielded with lead and one unshielded counter capable of handling 1,500 times the radiation intensity that had saturated the detectors in the first two Explorers; it also contained two scintillation counters, one to measure approximately the total energy and the other to count the incident corpuscular radiation. As the data came in, a plotting of the counting rates with reference to the earth's magnetic fields revealed clearly and unambiguously a radiation zone related to the lines of force of the geomagnetic field. Since the Advanced Research Projects Agency in the Department of Defense and the Atomic Energy Commission conducted in August a rocket test which produced a small high-altitude nuclear explosion, scientists had the additional benefit of data showing the effects of artificially introduced radiation of a known quantity and energy spectrum. By means of these observations and extrapolation, the State University of Iowa team mapped out the contours of the radiation zone and concluded that its structure must be more complex than they had suspected earlier. Deep space Pioneer probes launched by powerful rockets during the autumn of 1958 then enabled the experimenters to chart with some certainty what came to be known as the Van Allen radiation belts. This, the most significant scientific discovery achieved during the IGY, gave the United States a "first" in space exploration that wiped out most of the sting of having been second to the U.S.S.R. in putting a satellite into orbit.[16] Although American experts early realized that the first Sputniks might well have detected the existence of this phenomenon, had the Russians launched them at a higher angle of elevation to the earth, the American feat was none the less gratifying to the National Academy and the American public.

Each of the first three successful Explorer satellites had a short operating life, respectively under four months, less than three months, and just over ten weeks. From the six-inch *Vanguard I*, on the contrary, signals continued to come through clearly month after month despite the 2,460-mile apogee of the orbit; indeed receivers on the earth would be able to pick up the "beeps" for the next seven years. But because the grapefruit carried no instrumentation except the transmitters and two thermistors on its shell, the scientific information it could furnish at first looked meager—at least compared to the cosmic ray data deriving from the Explorers—and initially it appeared to be far less useful than that transmitted from the sophisticated Russian *Sputnik III*, 1958 Delta—even though signals from the latter were extremely difficult for American receivers to intercept.[17] Nevertheless, the little 1958 Beta proved more valuable than most people expected.

From study of the orbit, scientists at the Naval Research Laboratory and the Smithsonian Astrophysical Observatory were able to report some interesting discoveries: the earth is not a globe somewhat flattened at the poles, but is pear-shaped; the gravitational fields of the moon and the sun modify the orbit of earth satellites; the radiation pressure of light from the sun affects the movement of a satellite in its orbital path, and magnetic drag damps the rotational motion of metallic satellites. The repeated passages of the small artificial body enabled experts as time went on to determine the dimensions of the earth's equatorial and polar diameters, to demonstrate variations in atmospheric density with the rotation of the sun, and to show that the density of the upper atmosphere is far greater than formerly supposed. In 1961 when *Vanguard I* had been in orbit for three years, Hagen pointed out that if its life endured through the full solar cycle of eleven years, accurate estimates would be possible of the effect of atmospheric density upon drag and the satellite's length of life.[18] Although the mercury cell batteries ceased to function in June 1958, the solar cells continued to supply enough power to transmit signals to the Minitrack stations until 1965; thereafter optical tracking still permitted observation of orbital decay.[19] The probabilities are that the tiny object will remain in orbit for another 240 years.

With the highest apogee attained by any IGY satellite, *Vanguard I* achieved "a highly useful orbit," as the IGY summary report noted. Unlike the cylinder-shaped Explorer satellites, its spherical configuration saved it from tumbling and from developing propeller-like motions which hampered or prevented precision tracking. "The accelerations of 1958 Beta 2 correlate very well with occurrence of solar flares and the radio emission from the sun, and also show the 27-day solar revolution. This correlation, discovered by Luigi G. Jacchia, an eminent mathematician at the Smithsonian Astrophysical Observatory, is interpreted as arising from the heating of the atmosphere by solar radiation, causing the atmosphere to expand, thus increasing the density at high altitudes. Jacchia also discovered a bulging of the atmosphere, apparently from radiation heating, wherein the 600-km density level rises to about 950 km. The bulge follows the sub-solar point by approximately two hours." While 1958 Beta 2 was not the only American satellite to lend itself to calculation of a precise orbit, it furnished the material for a score of learned papers which like that of Jacchia, presented to the world new scientific knowledge.[20]

Hagen declared with justifiable pride that *Vanguard I* also contributed "firsts in the space program" by proving the effectiveness of solar cells as a source of power and by revealing "the peculiar and operationally annoying after-burning of solid propellant rockets." And, thanks to calibration of the crystal-controlled radio-frequency oscillator as a function of temperature, the tiny bird equipped only with two transmitters and two thermistors supplemented the data that came from the Sputniks and *Explorer I* and *III* about the extreme of heat and cold encountered under various conditions in a satellite or space vehicle.[21]

While the Academy's Technical Panel, Army and JPL scientists, and the Vanguard team were appraising the results of the launchings undertaken during the first half of 1958, the administrative arrangements under which the satellite program operated underwent change. In February, the Department of Defense transferred the direction of its share from the Assistant Secretary for Research and Engineering to the Advanced Research Projects Agency, and in April, when the bill to create the new civilian space agency came before Congress, the National Academy, in turn, prepared to adopt a somewhat different regime for pursuing research in space. During the preceding December, members of the TPESP had drafted a report to the National Committee recommending a long-term plan, calling first for experiments adapted to vehicles already under development, progressing step by step to "planetary and interplanetary investigations," and culminating in "manned space flight." The USNC executive committee had approved the proposals in January, and the 11 April 1958 issue of *Science* published them. Two months later, President Bronk's appointment of a Space Science Board, with Lloyd Berkner as chairman, to take charge of future planning, relieved the panel of one of its major responsibilities. The Working Group on Tracking and Computation and the WGII had already held their last meetings. The panel, after its session on 17 July, saw its own usefulness diminishing to the vanishing point, not only because passage of the National Aeronautics and Space Act on 29 July meant that NASA would soon set up its own scientific advisory staff, but also because impending changes in IGY management were likely sharply to reduce panel activities in channeling information to CSAGI.

At the CSAGI meeting in Moscow in August, the international committee announced that the IGY would run till 1960, but after 1958 in somewhat different guise: its official name would become the International Geophysical Cooperation and a body known as the Comité Internationale Géophysique, or CIG, would direct the program. CSAGI would go out of existence at the end of June 1959. Although CIG would devote to some fields of IGY interest less intensive effort than had prevailed during 1957–1958, clearly the satellite program would not suffer, for in October 1958, the International Council of Scientific Unions established a new international Committee on Space

Research (COSPAR) to deal with fundamental research in the celestial regions. Since the attempted launchings of two Explorers and one Vanguard satellite had failed during the late summer and fall, the extension of the IGY was especially gratifying to Americans.[22]

When NASA took over direction of Project Vanguard and the lunar probes in October 1958, the new agency absorbed most of NRL's Vanguard team and, through a Space Science Section at NASA headquarters, assumed most of the duties of the TPESP's working groups. As the Academy's Space Science Board handled other responsibilities formerly resting upon the panel, the TPESP met only once more, in July 1959, to make a last appraisal of the program. The Army Ballistic Missile Agency meanwhile remained in charge of Explorer development in Huntsville, at JPL, and at the Cape. At the beginning of the International Geophysical Cooperation in January 1959, Project Vanguard, still located physically at NRL, had four vehicles available for satellite flights; ABMA and the Jet Propulsion Laboratory were working to ready the powerful Juno II rocket for launching a 100-pound satellite.[23]

When *Vanguard II,* SLV–4, put 1959 Alpha into orbit on 17 February, it flew the cloud-cover experiment. The sensor system worked well, indicating in considerable detail the variations of the reflected earth radiation received by the satellite, but the data proved difficult to reduce because the satellite developed a large precession that caused it to move erratically, shifting its attitude relative to the earth.[24] Although the experimenters were therefore unable to make a complete mapping of the earth's cloud cover, the experience gained from the flight helped in designing and carrying out later meteorological experiments. Three unsuccessful American launching attempts, two Vanguard and one Explorer, followed before *Vanguard III,* 1959 Eta, began its orbit on 18 September.[25] As this was the last of the seven launch vehicles built under Navy aegis for the IGY, and as NASA decided not to commission more, Project Vanguard came to an official end shortly after this flight.[26]

Equipped with the Allegany Ballistic Laboratory's third-stage rocket with a fiberglass casing and nosecone, *Vanguard III* rose with a fifty-six pound payload, a weight made possible by the lightness of the fiberglass and by leaving the casing attached to the satellite during orbital flight instead of using a separation device. The twenty-inch sphere had a lower sector made of polished aluminum and an upper of fiberglass with a twenty-six-inch fiberglass tube projecting from it to support a magnetometer; it accommodated also the instruments for the solar x-ray and the Lyman-alpha experiments and the gauges for environmental study. The Lyman-alpha and solar x-ray experiments produced nothing useful because electrons in the Van Allen radiation belt swamped the ionization chambers with particles whose energies exceeded 150 keV. A seventy-day monitoring of temperature recorded changes ranging from about 40°C to −2°C; the average was about 20°C. No meteoritic penetration of the shell occurred, inasmuch as pressure readings remained constant. The impact rate of interplanetary matter, on the other hand, was highly variable, during one brief interval running as high as 1,900 an hour; a preliminary analysis put the influx of cosmic dust impinging upon the earth at about 10,000 tons a day.[27]

Of the more than 4,200 magnetometer signals received during 1959 Eta's eighty-four day flight, 2,872 were designated as a "prime data" set on the basis of quality and freedom from possible coded time errors. This set permitted charting the magnetic field with greater accuracy than ground measurements provided. The proton magnetometer, moreover, acted as a receiver for "whistler" signals in the 0.4- to 10-kc range and for a few "risers." The whistlers, very low frequency signals, came from dispersed lightning-produced ionospherics; the risers, according to some interpretations, were radiation from trapped particles. These observations enabled analysts to estimate electron densities above the F-peak of the ionosphere, to check theories of whistler propagation, and to study the conditions that allowed propagation of very low frequency signals from the troposphere to the satellite in the whistler mode. Since six periods of magnetic disturbance occurred during the operating life of 1959 Eta, measurements of the effects were attempted, but with generally inconclusive results. Greater disturbance was observable, however, at the northern and southern limits of the inner part of the outer Van Allen radiation belt than at magnetic latitudes greater than 25°.[28]

The finale of the IGY–IGC satellite program came with the launching of *Explorer VII* on 13 October 1959. A first Explorer VII had failed in July; eighteen months after the inception of the plan, the so-named "heavier payload" satellite with seventy pounds of instruments for six experiments was at last in orbit. Enormous effort had gone into design of the multiple package. As early as April 1958 representatives from the Army Ballistic Missile Agency had met with NRL experts in order, in Roger Easton's phrase, "to permit ABMA to learn as much about satellites as possible in the least possible time." As the heavily laden bird was to accommodate a 20-mc transmitter as well as two 108-mc transmitters, the layout had posed "a stiff problem" in arranging the telemetering equipment and the complex scientific instrumentation. The Academy's panel, deviating from its earlier decision to sponsor no new experiments for IGY–IGC flights, had requested the inclusion of instruments designed by Hermann LaGow of NRL to detect, by means of cadmium-sulphide photosensitive cells, micrometeoroid erosion and penetration. So the second *Explorer VII* [29] carried the NRL solar x-ray, Lyman-alpha, and micrometeor detection instruments, an elaboration of Van

Allen's and Ludwig's earlier cosmic radiation apparatus, the equipment for Pomerantz's and Groetzinger's heavy nuclei experiment, and, sixth, Suomi's sensors for measurement of the earth's radiation balance which, after the failure of the Vanguard SLV–6 launching in June, were redesigned to fit into the Explorer package.[30]

The results obtained from several of the experiments were disappointing. Again Friedman's ion chambers were flooded by radiation electrons. This time, however, by plotting the latitude, longitude, and altitude of the points at which the saturation occurred, it was possible to identify trapped radiation as the cause and to plan an experiment to be flown in an orbit that would not enter the radiation belts. Carried out in June 1960, that scheme successfully recorded ultraviolet and solar x-ray radiation from a solar flare during its onset and development. In the micrometeoroid detection experiment in *Explorer VII*, 1959 Iota, one of the three Cds cells was damaged and desensitized during the launching; another, designed primarily for calibration of the sunlight penetrating it, was relatively insensitive; and the signals received from the third cell were of a character that precluded reducing the data to satisfactory form. For a time, defeat also threatened the acquisition of usable readings of the flux of heavy primary cosmic ray nuclei, for the ion chamber encountered interference from the solar radiation experiment, and the circuitry associated with one channel early underwent a change in mode of operation. Although the consequent rather fragmentary data were hard to translate, study of recordings made over a six-month period around the world in the northern hemisphere eventually permitted plottings of the integral energy spectrum and of changes in its shape.[31]

The additional information collected about the Van Allen radiation belts, on the other hand, quickly supplemented that assembled from earlier satellites. Signals recorded a number of solar-terrestrial-coupled events—the arrival of solar protons following their acceleration in a solar flare, for example, and a polar cap display marked by increased ionospheric absorption inside the auroral zone. From observations made during a severe geomagnetic storm on 29 November 1959 an hypothesis evolved that, when solar plasma encounters the earth's magnetosphere, a distortion of the magnetic field occurs in fashion that causes particles normally trapped in the outer radiation belt to be "dumped" out of the belt so as to interact with the atmosphere at altitudes below the mirror points and to spread to lower latitudes. The direct correlation shown by *Explorer VII* between occurrences in the radiation belts and auroral activity in the high atmosphere supported the "dumping" theory. The delineation of zones of geomagnetically trapped, high-energy particles, to be sure, left many unknowns, but it widened knowledge of the "population identity" and energy spectrum of the trapped particles. And it gave clues to the mechanism of trapping, helped explain the behavior of the belts during solar and interplanetary disturbances, and clarified the relationships between terrestrial manifestations of solar disturbances and activity in the belts.[32]

The measurements of the earth's radiation balance were also significant, albeit less complete than Suomi hoped for. Even so, with as many as 432,000 separate measurements made in a single month, the accumulation gave meteorologists more than they could work through in the next seven years. As *Explorer VII* carried no storage unit, data reduction was difficult. Scientists first undertook analysis of the earth's radiation losses, leaving till later the computation of the gains from the sun in the earth's heat budget; the findings on gains were only beginning to emerge at the end of 1965. As redesigned for *Explorer VII*, the essential instrumentation for this experiment consisted of glass-coated bead thermistors making contact with the sensors, two spheres, and four hemispherical bolometers, that is, electrical devices that register minute quantities of radiant heat. The spheres, one black-coated and one fitted with a shade to protect it from direct sunlight, were mounted on the spin axis of the satellite. The four bolometers were placed in the satellite's equatorial plane close to, but thermally insulated from, a mirror so coated as to have high resistance in the ultraviolet. One hemisphere, coated white, was more sensitive to terrestrial than to solar radiation; two black-coated hemispheres responded about equally to solar and terrestrial radiation, while the fourth, coated with gold, responded chiefly to solar.

Although the lack of a data storage unit prevented a synoptic mapping of fields of radiation outgoing from the earth, study of the measurements indicated that patterns of a large-scale outward flux of radiation exist and are related to large-scale features of the weather; cloud cover and circulation patterns control the earth's loss of radiation; and within the atmosphere a pronounced vertical divergence of net long-wave radiation occurs. Further study permitted meteorologists in the course of time to estimate the heating and cooling of the atmosphere and to make a beginning on gauging the role of differential cooling in supplying atmospheric energy.[33]

At the official termination of the IGC on 31 December 1959, some three weeks after *Explorer VII* had ceased to relay signals to the earth, reduction of the telemetered data had not progressed far; meaningful interpretations of all the findings would take years. Indeed, in 1967 experimenters would still be examining the results of IGY–IGC satellite flights. But by 1960 the richness of the scientific harvest from the satellite program was already manifest to the scientific world. "Space science," a little ruefully defined by an academician as "any scientific inquiry that NASA will pay for," had come into its own.

An ad hoc committee met at the Boulder Laboratories of the National Bureau of Standards, Boulder, Colorado, on 16 December 1959 to discuss continued operation of the U.S. World Data Centers. These centers— World Data Center A for Airglow and Ionosphere at Boulder and the one for Solar Activity at the University of Colorado—had been set up during the IGY for collection and storage of research data. Present at the meeting (left to right): John Lyman, NSF; John R. Winkler, University of Minnesota; Homer E. Newell, Jr., NASA; Alan H. Shapley, NBS, convenor of the meeting and Vice-chairman of the United States National Committee for the IGY; Hugh Odishaw, NAS, and Executive Director of the National Committee for the IGY; and Pembroke Hart, NAS.

THE FINAL ACCOUNTING

For billions of years, planet Earth had been accompanied by one celestial body in its journey around the sun; in two years at the end of the 1950s nineteen new satellites with lifetimes ranging from a few months to more than a hundred years were launched.[1] In evaluating the scientific accomplishments of the IGY satellite program the layman must feel a sense of bewilderment. The terms used to describe experimental goals seem esoteric, and the significance of the results and their connection with phenomena on earth are even more difficult to grasp. In many respects the layman's uncertainty is shared by scientists. The voyages of the IGY satellites were true voyages of discovery. Many of the findings were completely unexpected and often derived from experiments aimed at completely different phenomena. Areas of investigations which had hitherto commanded little attention were thrust into the forefront of scientific concern, and scientists found themselves learning a new vocabulary and confronting new problems.

If it is possible to summarize the findings of the satellite program in any intelligible way, two statements may suffice: the studies revealed the extent of the earth's influence in space, and at the same time they showed just how little we understood of the environment of the earth. Analysis of the motion of *Vanguard I,* for example, by establishing the fact that the earth is not spherical but rather has a bulge, disclosed unsuspected stress deep within the earth. Analysis of the drag exerted by the atmosphere on *Vanguard I* proved the atmosphere to be far more extensive and variable in extent than believed before. *Explorer I* revealed the complex region of charged particles and magnetic fields surrounding the earth. Since the end of the IGY scientists have studied the nature and extent of the Van Allen belts in detail, but physicists themselves as yet understand very little about the processes involved in the phenomena and their relationship to the more familiar atmosphere below.

If much of the nonscientific public tended to dismiss the results of the other IGY experiments as fragmentary and accordingly inconsequential, scientists attributed some importance to all data acquired in this new realm of research. In their endeavor to build up a body of knowledge about the earth's environment, they had to rely to a considerable degree on patiently piecing together scraps of information about extraterrestrial space. The IGY satellites had begun to pile up evidence about happenings beyond the ionosphere; further investigations promised to establish causal relationships among obscure phenomena. The failure of the first two attempts to measure variations in Lyman-alpha intensity during a satellite's orbit had in itself proved useful, first by the indication of strength of the radiated electrons that flooded the ionization chambers and then, after the plotting of the locations at which the saturation occurred, by showing what orbital paths a satellite must avoid when seeking data on ultraviolet radiation. Thus a third try brought success.

In appraising the satellite program, it is important to realize how much the success of its scientific phases owed to the efforts of the Vanguard team and other NRL scientists intimately associated with the project. It was they who worked out the principles and methods of thermal control, devised electronic equipment of exceptional reliability, and tested every mechanism in the packages of experiments. Whether flown in Explorers or Vanguards, the experimenters' apparatus had a dependability that stemmed in no small measure from the work lavished upon it at the Laboratory. That several experiments failed to produce the kind of data scientists hoped to obtain was never due to weaknesses in the construction of the instruments or in the telemetry, or to the integration of parts into the satellite structure. The adaptations JPL made to fit experiments designed for Vanguard into Explorers were skillful, but the basic problems were already solved before JPL engineers undertook the assignment.

What Project Vanguard, as the hardware part of the program, contributed to space exploration is widely misunderstood, doubtless partly because questions about the might-have-beens obtrude themselves. Some ask, for example, if the choice of launching system had fallen on the Army Orbiter, and if, as is highly probable, the United States had therefore been first to put up a satellite, would not the gains in American prestige have outstripped any benefits that ultimately accrued from developing a vehicle employing novel features of design and engineering and equipped with miniaturized instruments? Or, again, might not long-term progress have been faster if the National Security Council had assigned the program a top priority and the Department of Defense had then authorized production of the vehicle on a crash basis, or if DoD policy-makers had decided to risk the perils of service rivalries and made the Army responsible for the launcher, the Navy for the instrumentation, including the telemetry and tracking systems? Others, conversely, raise a very different question: wasn't the jolt to American pride in American technological superiority to all other nations a salutary blow? If the United States had beaten the Russian time-table, would not that success have perpetuated our national complacency and delayed disastrously the reexamination of our educational system and the teaching of mathematics and science in our schools? All these speculations tend to becloud judgments on Project Vanguard.

In the opinion of well-informed people, a major handicap that beset Project Vanguard from the start was its relegation to a status secondary to the ballistic missile program. The White House decree that procurement and testing of the satellite vehicle must not interfere with top-priority military projects left the Laboratory and its prime contractor more than once in an awkward position; certainly Vanguard's low priority slowed the flow of money and delayed progress on the vehicle in less obvious ways. A number of members of the Academy's IGY committees and panels and some of the Laboratory's staff named three other factors that, in their view, hampered Project Vanguard: first, the Martin Company's taking on the Titan contract and a consequent dilution of the contractor's interest in the satellite launcher; second, the appointment of a radio astronomer instead of an experienced rocket engineer as project director; and, third, the blaze of publicity in which the Vanguard teams had to operate, subjecting them constantly to Sunday morning quarterbacking and inflating their every setback to the proportions of a disaster born of ineptitude or negligence.

Time has disposed of the once frequently voiced belief that John Hagen permitted things to get out of hand, for all the men deeply involved in the project have come to see that his handling of Vanguard's many-faceted problems preserved a necessary balance between the scientific and technical aspects of the program. Whether the other two conditions cited as obstacles were indeed the source of serious trouble may be debated. Had they not obtained, one can only conjecture whether Vanguard engineers would have bettered their score of placing three satellites in orbit out of eleven tries. In any case, insofar as these factors affected Project Vanguard adversely, the damage done was chiefly to the time schedule rather than to the ultimate performance. The only valid estimate of the degree of success or failure in the undertaking must rest upon what the Naval Research Laboratory and the Martin Company did accomplish, not upon what a different regime might have achieved, or what other circumstances would have permitted under Navy aegis.

Some critics have contended that NRL's approach to design of the launching system was intrinsically faulty: in attempting to employ a first-stage rocket of marginal power, the originators of the plan had to rely on over-elaboration, compensating for minimum engine thrust by recourse to lightweight materials, miniaturization, and precision work, instead of using off-the-shelf components that otherwise would have served as well and at far less cost in time. Yet these very limitations led to many of Vanguard's most valuable contributions to the art of rocketry and miniaturization of electronic components. The engineering feat of designing, constructing, and testing within thirty months a vehicle that could and did launch an earth satellite was in itself extraordinary, especially at a time when the art of rocketry was still in adolescence. Wernher von Braun, chief architect of the Redstone, Jupiter-C, and Juno rockets, called it a miracle. Whereas the "man on the street" today is likely to look blank at mention of Project Vanguard or else identify it as "that thing that blew up," James Bridger of NRL, when asked how he ranked it, replied: "I'd call it 300 percent successful. Our job was to get one satellite into orbit during the IGY; we put up three." Ultimately more important were the technological innovations introduced in Vanguard—advances in design and in the use of materials that have influenced rocket engineering for a decade. While one school of thought has contended that Vanguard's mission did not extend to the development of models for post-IGY satellite launchers and that such work cannot properly count in any assessment of Vanguard achievements, most space engineers consider it an important entry on the asset side of the ledger.

Of the major innovations, the use of miniaturized circuits and batteries was one of the most valuable. The solar cells developed by the Signal Engineering Laboratories and so placed by Vanguard experts on the satellite shell as not to interfere with the functioning of the internal instrumentation set a new standard of efficiency and accounted for the long operating life of *Vanguard I*. Later satellites produced by NASA have similarly employed solar power. The use of unsymmetrical dimethylhydrazine as fuel in the second-stage rocket was another significant new departure, as indeed was much of the design of the Aerojet rocket. Impressed by the economy and utility both of the second stage and the Grand Central Rocket Company's third-stage rocket, the Air Force bought and used them in its Thor-Able booster. The fiberglass-encased third-stage motor devised by the Allegany Ballistic Laboratory and flown in *Vanguard III* was in turn a pioneering development; with a mass ratio of 0.91, it achieved a specific impulse of 251 seconds, a notable record for a solid-propellant rocket. Furthermore, the use of a "strapped-down" gyro platform, the rotatable exhaust jets of the first-stage turbopump which ensured efficient roll control, and the C-band radar beacon antenna employed on the Thor-Able vehicle all originated with Vanguard. Nor can a fair appraisal overlook the importance to the emerging

254

spacecraft industry of the elaborate and original techniques the Martin Company developed to define and solve such problems as optimization of the trajectory and preflight predictions of the vehicle's performance.[2]

The most striking evidence of Vanguard's rich legacy to the design of later spacecraft came with the appearance of the Delta launcher, built to NASA specifications by the Douglas Aircraft Company and first flown in May 1960. The Delta second stage is the Vanguard with a few modifications —repackaged electronic components, a stainless steel instead of an aluminum thrust chamber, and a new radio guidance system. Delta's third stage is a replica of the Allegany Ballistic Laboratory's rocket used in the last Vanguard. The most versatile and reliable of all American space vehicles, Delta between 1960 and the spring of 1968 put into orbit fifty-two satellites ranging in type from weather satellites to orbiting solar observatories. "An equally important story," Milton Rosen declared, "can be told about the Stadan network, a living descendant of Minitrack, through which flows day after day, week after week, year after year, the major portion of NASA's scientific output." [3]

Another product of Vanguard experience was a new method of budget forecasting and cost reporting specially adapted to contracts in which research and development features made expenditures peculiarly difficult to estimate in advance and hard to keep track of in orderly fashion. This scheme to enable the project director to report on contractors' progress and financial needs was inaugurated by Tom Jenkins in the autumn of 1956. Under its provisions, the forms sent out by the comptroller to the Martin Company, and to the firms with whom the Laboratory had direct contracts, contained columns specifying the breakdown of the information required monthly. These forms quickly proved helpful to company finance officers and a boon to the government. NASA made use of the system, and, after Jenkins drafted a study explaining its workings, it was adopted by other agencies.[4]

Finally, the "development-testing philosophy" worked out at the Naval Research Laboratory constituted a contribution to the program that NASA recognized as basic and has acted upon consistently. The Laboratory applied it both to the vehicle and the payload. After suffering from proven charges of careless workmanship in the fabrication of the vehicle, the Martin Company in 1958 adopted the standards held up as a must by Commander Berg and, in order to proclaim the meticulous exactitude thereafter demanded of its manufacturing and inspection units, the company used as its advertising slogan: "The margin of error is zero." While static and flight testing in the field differed little in most respects from the routines observed by the Air Force and the Army, rigid calibrations of the first-stage engine during static firings and precise procedures of loading the propellant enabled Vanguard engineers to keep first-stage propellant outage at a low level without installing an automatic fuel utilization system.[5] After the failure to get the TV–3 backup into orbit, the Martin and Laboratory staffs always undertook a thorough analysis of the performance of every part of every vehicle fired, whether the flight was successful or failed. In this way engineers were able to identify and correct deficiencies and thus greatly improve the reliability of each successive vehicle.

The techniques of testing the satellite's instrumentation were still more exacting. Assembling equipment that would simulate the conditions to be encountered in the vacuum of space and developing processes that would reveal the nature of weaknesses in experimenters' apparatus required scientific knowledge and expert craftsmanship. From the shake table used in measuring vibration resistance to the vacuum tank and the gauges employed in registering temperature changes in the satellite placed within it, every test device was carefully constructed and operated. Testimony to the quality and thoroughness of the procedures lay in the fact that all the instrumentation flown in IGY satellites functioned properly.

All in all, the record is clear. Project Vanguard justified the faith of its supporters not only by putting instrumented satellites into orbit during the IGY but by developing a vehicle with "growth potential," and, in the process, by advancing the art at a cost in money that, in 1961, looked incredibly small to experts. The one black mark against it is that it did not "get thar fustest." Homer Newell's summary judgment ran: "A failure? Vanguard was a resounding success!" [6]

The oblivion to which most Americans consigned Project Vanguard at the end of the 1950s no longer distresses the men who gave three years of their lives and ate their hearts out in the endeavor to make Vanguard a landmark on the unending road to new scientific knowledge. They know that it was and is a landmark. Scientists at the National Academy and experienced leaders at NASA acknowledge it as a progenitor of all American space exploration today. "The overall scientific program developed for use with the Vanguard launching system," stated the satellite panel toward the end of its life, "has made possible the total program of space vehicle instrumentation, observation, and data reduction carried out under IGY auspices. Additionally, it has provided the original basis of the present expanding program of scientific experiments for space research for the United States." [7] True, it did not produce the first artificial satellite to circle the earth; true also, the initially rejected launching system nurtured by the Army Ballistic Missile Agency and refined by JPL put up the apparatus that netted the most valuable data collected during the IGY. But those facts have become largely details of history. Recognition of the Explorer achievement should not denigrate Vanguard's, just as Vanguard's cannot detract from Explorer's. To the pure scientist all that matters is that a new research tool became available and its proven utility has ensured its continuing use. To thousands of imaginative Americans of the 1960s, the expansion of knowledge of interplanetary and solar space represents a creative undertaking in its own realm as inspiring as the work of the cathedral builders of the Middle Ages.

NOTES

Chapter 1

[1] *Collected Works of K. E. Tsiolkovskiy*, vol. 2 (*Reactive Flying Machines*), A. A. Blagonravov, ed., NASA Technical Translation TT F-237 (Washington, D.C.: National Aeronautics and Space Administration, 1965), translation of *K. E. Tsiolkovskiy: Sobraniye Sochineniy*, Tom II, *Reacktivnyye Letatel'-nyye Apparaty* (Moscow: Izdatel'stvo Akademii Nauk SSR [USSR Academy of Sciences Publishing House], 1954); Hermann Oberth, *Die Rakete zu den Planetenräumen* (Munich and Berlin: Oldenbourg, 1923), and "From My Life," *Astronautics*, vol. 4, no. 6 (June 1959), pp. 38, 100, 106. Oberth built a rocket for a German movie, but it was never flown.

[2] Willy Ley, *Rockets, Missiles and Space Travel*, 3 rev. ed. (New York: Viking Press, 1961), p. 323.

[3] Milton Lehman, *This High Man: The Life of Robert H. Goddard* (New York: Farrar, Straus & Co., 1963). More on Goddard's pioneering work will be found in his papers, edited by Mrs. Esther Goddard and G. Edward Pendray in three volumes, *The Papers of Robert H. Goddard* (New York: McGraw-Hill, 1970).

[4] Interviews with Dr. Lloyd Berkner, 20 Mar 1965, and Dr. Harvey Hall, 6 June 1967. As nearly every scientist who participated in the American satellite program was a Ph.D., the academic label is hereafter omitted.

[5] Milton Rosen, *The Viking Rocket Story* (New York: Harper & Bros., 1955), p. 18; interview with Edward O. Hulburt, Director of Research at the Naval Research Laboratory, 1945-1955, 8 Apr 1964. I. B. Holley, Jr., an Army Air Forces officer and historian stationed at Wright Field during 1945, recalls the briefing and the arrival of the German engineers.

[6] Rosen, *op. cit.*, pp. 22-23; see also "Historical Origins of Echo I," pp. 4-5, 15, interview of William J. O'Sullivan, by Edward Morse, 23 Aug 1964, mimeo, NASA historical files (hereafter cited as NHF).

[7] The chronology of NRL-Viking-Vanguard developments derives from an official NRL summary, prepared in October 1957 and entitled "Decisions on Prior Programs Affecting Project Vanguard and Ballistic Missile Development," which is enclosure 1 to NRL letter 4100-227 (hereafter cited as encl 1, NRL 4100-227).

[8] Harvey Hall, memo for file, ONR: 405, 29 Nov. 1957, copy in NHF; interview Harvey Hall, 6 June 1967; Wernher von Braun, "From Small Beginnings," in Kenneth W. Gatland, et al., *Project Satellite* (London: Wingate, 1958), pp. 47-49; von Braun statement in F. Zwicky, "Report on Certain Phases of War Research in Germany," pp. 38-42, Headquarters, U.S. Air Materiel Command, January 1947.

[9] Hall, ONR:405, pp. 2-3.

[10] Negative evidence, namely the lack of memos in NRL files regarding the ESV proposal, substantiates this statement.

[11] H. Hall memo, ONR:405, p. 4.

[12] Project RAND, *Preliminary Design of an Experimental World-Circling Spaceship*, May 1946, cited by R. Cargill Hall, "Early U.S. Satellite Proposals," in *The History of Rocket Technology: Essays on Research, Development, and Utility*, Eugene M. Emme, ed. (Detroit: Wayne State Univ. Press, 1964), pp. 75-79 (hereafter cited as *History of Rocket Technology*).

[13] U.S. Congress, Senate, Committee on the Armed Services, Preparedness Investigating Subcommittee, Hearings: *Inquiry into Satellite and Missile Programs*, part 1, 85th Cong., 1st and 2d Sess., November 1957 and December 1958, p. 283; Theodore von Kármán, *Toward*

New Horizons: A Report to General of the Army H. H. Arnold, Submitted on Behalf of the A.A.F. Scientific Advisory Group, Army Air Forces report, 15 Dec 1945, p. 56. Von Kármán's autobiography (written with Lee Edson), *The Wind and Beyond,* was published in 1967 by Little, Brown & Co., Boston.

[14] H. Hall, memo, ONR:405, pp. 4–5.

[15] Quoted by R. Cargill Hall in *History of Rocket Technology,* p. 79.

[16] See n. 14; *First Annual Report of the Secretary of Defense,* 1948, p. 129; memo, RDB 57–3086, appendix 4; R. Cargill Hall in *History of Rocket Technology,* pp. 87–88, n. 85.

[17] H. Hall memo, ONR:405, pp. 2, 4–5; "Historical Lessons from Research and Development: The Case of Navaho," citing Francis Bello, "The Early Space Age," *Fortune,* vol. 16, no. 1 (July 1959), typescript in NHF.

[18] R. Cargill Hall in *History of Rocket Technology,* p. 90.

[19] See, for example, James A. Van Allen, L. W. Frazer, and J. F. R. Lloyd, "Aerobee Sounding Rocket—A New Vehicle for Research in the Upper Atmosphere," *Science,* vol. 108, no. 2818 (31 Dec 1948), p. 746.

[20] Rosen, *The Viking Rocket Story,* pp. 10, 35–39.

[21] *Ibid.,* pp. 239–241; NRL 4100–277, encl 1 and encl 3, "Significant Results of Vanguard and Previous High Altitude Rocket Projects." See also, Rocket Research Report, No. 6, "Conversion of Viking into a Guided Missile," and 11, "A Phase Comparison Guidance System for Viking," U.S. Navy, Naval Research Laboratory Report (hereafter cited as NRL Report) 3829, 1 Apr 1951, and 3982, 5 May 1952.

[22] NRL 4100–227, encl 4, "Funding History of Related Projects Synopsis FY 1945–1958."

[23] NRL 4100–227, encl 1, 3, and 4; NRL Reports 4576, 4727, 4757, and 4899; William R. Corliss, draft, "The Evolution of STADAN," p. 1, 1 June 1966. This draft was later published as *The Evolution of the Satellite Tracking and Data Acquisition Network (STADAN),* Goddard Historical Note No. 3 (Greenbelt: Goddard Space Flight Center, January 1967).

[24] 64 Stat. 149, 10 May 1950; interview, Alan T. Waterman, Director NSF, 1950–1963, November 1966.

[25] E.g., National Academy of Sciences, Report, Committee on Upper Atmosphere Rocket Research, no. 33, October 1952, copy in NHF.

[26] Clayton S. White and Otis O. Benson, Jr. eds., *Physics and Medicine of the Upper Atmosphere: A Study of the Aeropause* (Albuquerque: University of New Mexico Press, 1952); interview, Joseph Kaplan, 14 Feb 1968.

[27] Interview, Rosen, 22 Nov 1966.

[28] J. R. Pierce, under the pseudonym of J. J. Coupling, "Don't Write: Telegraph!" in *Astounding Science Fiction* (March 1952), vol. 49, pp. 82–96.

[29] American Rocket Society Space Flight Committee, "On the Utility of an Artificial Unmanned Earth Satellite," *Jet Propulsion,* vol. 25, no. 2 (February 1966), pp. 71–73.

[30] Shirley Thomas, *Men of Space,* vol. 2 (Philadelphia: Chilton Book Co., 1960–1968, 8 vols.), pp. 190–194; interview, Lloyd Berkner, 20 Mar 1965. A copy of Grosse's report to President Truman, "Report of the Present Status of the Satellite Problem," 25 Aug 1953, is in NHF.

[31] *Collier's,* vol. 129, no. 12, and vol. 130, nos. 16 and 17 (22 Mar and 18 and 25 Oct 1952); Ley, *Rockets, Missiles and Space Travel,* pp. 324–325; interview, Kaplan, 14 Feb 1968.

[32] Ley, *op. cit.,* p. 325; Harry Wexler, "Observing the Weather from a Satellite Vehicle," *Journal of the British Interplanetary Society,* vol. 13, no. 5 (September 1954), pp. 269–276; Singer, "Studies of a Minimum Orbited Unmanned Satellite of the Earth (MOUSE)," Part I, Geophysical and Astrophysical Applications, *Astronautics,* vol. 1, pp. 171–184. See also E. Nelson Hayes, "The Smithsonian's Satellite-Tracking Program: Its History and Organization," *Annual Report . . . Smithsonian Institution . . . 1961,* pp. 276–277, 1962,

[33] See n. 18; Wernher von Braun, "The Redstone, Jupiter and Juno," in *History of Rocket Technology,* pp. 109–110.

[34] Wernher von Braun, "A Minimum Satellite Vehicle, Based on Components Available from Missile Developments of the Army Ordnance Corps," Guided Missile Development Division, Ordnance Missile Laboratories, Redstone Arsenal, Huntsville, Alabama, 15 Sept 1954.

[35] Ltr, Whipple to James Van Allen, 10 June 1955, Correspondence File, 1–9, National Academy of Sciences, IGY files (hereafter cited as

NAS–IGY files) ; interview, Alan T. Waterman, 6 Oct 1966.

[36] R. Cargill Hall, "Origins and Development of the Vanguard and Explorer Satellite Programs," draft copy, October 1963 with annotation by Homer J. Stewart of JPL, NHF; memo, Dr. E. R. Piore, ONR, 21 Dec 1954, NHF; Commander George W. Hoover, "Why an Earth Satellite," prepared for presentation to the Engineers' Society of Milwaukee, 22 Feb 1956; Whipple notes on draft of this chapter.

[37] Interviews, Lloyd Berkner, 26 Mar 1965, and J. Wallace Joyce, 19 July 1967; "The Scientific Earth Satellite," remarks by Dr. J. W. Joyce, Head, Office for the International Geophysical Year, National Science Foundation, before the 27th Annual Meeting of the Society of Exploration Geophysicists, 12 Nov 1957, NHF. See also Jay Holmes, *America on the Moon: The Enterprise of the Sixties* (Philadelphia: J. B. Lippincott Co., 1962) , pp. 43–48.

[38] "Catalog of Data in the World Data Centers," vol. 36 of *Annals of the International Geophysical Year* (New York: Pergamon Press, 1964), p. vi.

[39] Joseph Kaplan, "The United States Program for the International Geophysical Year," in National Academy of Sciences and National Research Council *News Report*, vol. 4, no. 2, pp. 17–20, 1954.

[40] *Congressional Record*, 81st Cong., 2d Sess., pp. 2401–2407, 2411; 67 Stat. 488, 8 Aug 1953; Minutes of the 1st Meeting USNC, NAS-IGY Files; *The National Science Foundation, A General Review of its First 15 Years*, Report of the Science Policy Research Division, Legislative Reference Service of the Library of Congress, to the Subcommittee on Science, Research and Development of the House Committee on Science and Astronautics, 89th Cong., 1st Sess. (Committee print) , p. 33 (hereafter cited as *National Science Foundation First 15 Years*); National Science Foundation *Annual Report*, 1956, p. 26. See also ltr, William W. Rubey, Chairman, National Research Council of the National Academy of Sciences, to Alan T. Waterman, Director of the National Science Foundation, 25 Nov 1953, Berkner papers, NAS-IGY files.

[41] Interviews, Lloyd Berkner, 20 Mar 1965, Hugh Odishaw, 21 May 1965, and Homer Newell, 18 July 1966; Jay Holmes, *America on the Moon*, pp. 44–49.

[42] Quoted in appendix 4, memo RDB 57–3086. See also John P. Hagen, "The Viking and the Vanguard," *Technology and Culture*, vol. 4, no. 4 (fall 1963), pp. 435–437.

Chapter 2

[1] Minutes of 1st Meeting of the Technical Panel on Rocketry, 22 Jan 1955; interviews, Hugh Odishaw, 8 July 1966, Homer Newell, Jr., 18 July 1966, and Athelstan Spilhaus, 18 Dec 1967.

[2] See above, pp. 15–16.

[3] Interview, Milton Rosen, 26 Feb 1964; NRL ltr, 4100–227, encl 1, 1955; ltr, Rosen to R. Cargill Hall, 28 Aug 1963, in re Hall's draft of "Origins . . . of the Vanguard and Explorer Satellite Programs," copy NHF.

[4] Report of Subcommittee on Long-Playing Rocket, attached to Minutes, 2d Meeting, Technical Panel on Rocketry, 11 Feb 1955; Minutes, 3d Meeting 9 Mar 1955; interview, Athelstan F. Spilhaus, 18 Dec 1967.

[5] Interviews, George Derbyshire, 15 Dec 1967, Athelstan F. Spilhaus, 18 Dec 1967, and Hugh Odishaw, 26 Dec 1967.

[6] Ltr, Kaplan to Waterman and Bronk, 14 Mar 1955, IGY correspondence file, 1–9.

[7] Ltr, Waterman to C. McL. Green, 15 July 1965, listing entries from his diary from March to August 1955; memo RD 57–3086, appendix 5, 25 Oct 1957 (S) ; Homer J. Stewart notes on draft manuscript, R. Cargill Hall, "Origins . . . of the Vanguard and Explorer Satellite Programs," March 1964; interview, Milton Rosen, 29 Aug 1966; NRL Report, "A Scientific Satellite Program," 13 Apr 1955.

[8] Ltr, Kaplan to Waterman, 6 May 1955, USNC Executive Committee, Correspondence files; interviews, Waterman, 10 June 1965, and General Dwight D. Eisenhower, 6 Nov 1966; Minutes, 8th Meeting USNC, 18 May 1955.

[9] Unless otherwise noted, the sources for the discussion in this paragraph and the rest of this chapter derive from privileged documents and interviews.

[10] See White House release, "Statement by the President," 9 Oct 1957, p. 2.

[11] *Ibid.*

[12] Memo, Deputy Secretary/Defense to Secretary/Army, Secretary/Navy, Secretary/Air Force, 8 June 1955; interview Alan T. Water-

man, 12 June 1965; Homer J. Stewart, notes on R. Cargill Hall's draft manuscript "Origins . . . of the Vanguard and Explorer Satellite Programs."

[13] RD 263/9, appendixes E and F, 4 August 1955; interview, Hugh Odishaw, Executive Director, USNC, 1 Sept 1966; Milton Rosen diary, 7 July 1955.

[14] Ltrs, Kaplan to Waterman, 6 June, to Odishaw, 18 July, and to Sydney Chapman, 26 July 1955; memo, Peary to Odishaw, 13 June 1955; *National Science Foundation First 15 Years* (see chapter 1, n. 40), p. 33; Waterman, diary entries, June and July, and ltr to Quarles, 17 June 1955.

[15] Interview, Alan T. Waterman, 12 June 1965.

[16] Interview, Alan T. Waterman, 10 June 1965; Secretary of Defense and National Academy and Science Foundation press releases, 29 July 1955.

[17] New York *Herald Tribune*, 3 Aug 1955, ltr, Kaplan to Waterman, 12 Aug 1955, IGY correspondence files.

Chapter 3

[1] Report of the Committee on Special Capabilities to Assistant Secretary of Defense for Research and Development, 4 Aug 1955, RDB 55–1607A, B–3.

[2] Homer J. Stewart, notes on draft manuscript, R. Cargill Hall, "Origins . . . of the Vanguard and Explorer Satellite Programs," August 1963.

[3] Memo RDB 55–1607A, C, 1–4.

[4] See pp. 28–29.

[5] NRL Memo Report 487, "A Scientific Satellite Program," 5 July 1955. The Laboratory had worked out the design of both the M–10 and the M–15 when pursuing the guided missile reentry problem for the Air Force in 1954. See p. 12.

[6] NRL Memo Report 487, 5 July 1955. See also pp. 121–123.

[7] NRL Memo Report 487, 5 July 1955. See also William R. Corliss, *The Evolution of STADAN*, Goddard Historical Note No. 3 (Greenbelt: Goddard Space Flight Center, January 1967), pp. 17–18.

[8] See Kurt R. Stehling, *Project Vanguard* (Garden City, New York: Doubleday & Co., Inc., 1961), p. 53.

[9] Tape-recorded interview, Homer J. Stewart with Eugene Emme, 1960.

[10] RDB 55–1607; ltr, as dictated to Thomas Lauritsen by his father Charles Lauritsen, to C. McL. Green, 29 May 1967.

[11] Interviews, Richard Porter, 1 Feb 1968, and George Clement, 8 Mar 1968.

[12] H. J. Stewart's notes to C. McL. Green, 14 Mar 1968.

[13] RDB 55–1607A and RDB 55–1607.

[14] Clifford C. Furnas, "Why Vanguard?" *Life*, vol. 43, no. 17 (21 Oct 1957), pp. 22–25.

[15] RDB 55–1607; Homer J. Stewart's notes to C. McL. Green, 14 Mar 1968.

[16] "Project Vanguard, a Scientific Earth Satellite Program for the International Geophysical Year," Report to the Committee on Appropriations, U.S. House of Representatives, by Surveys and Investigations Staff, in U.S. Congress, House of Representatives, Committee on Appropriations, Subcommittee on Department of Defense Appropriations, *Hearings on Department of Defense Appropriations for 1960*, part 6, 86th Cong., 1st Sess., 14 Apr 1959, pp. 62f. See pp. 8–9 and 16–18.

[17] Notes, Homer J. Stewart to C. McL. Green, 11 Oct 1966.

[18] See n. 16.

[19] See n. 17 and chapter 12.

[20] Memo, Rosen to C. McL. Green, 12 Oct 1966; memo, Director NRL to Chief of Naval Research, 15 Aug 1955, S–1000–26/55 LP, Ser 2307.

[21] Memo, Rosen to Director, NRL, 23 Aug 1955, sub: Addendum to NRL Memo Report 487 of July 1955.

[22] Memo, Rosen to C. McL. Green, 12 Oct 1966.

[23] Encls 1–4 to Rosen memo cited in n. 21.

[24] Notes, H. J. Stewart to C. McL. Green, 11 Oct 1966.

[25] Encl 2, "Chronology of Significant Events in the Vanguard Program" to Hagen ltr, NRL 4100–227, 10 Oct 1957; memo, Dpty Sec/Def to Secs/Army, Navy, and Air Force, Sec/Navy Cont. S–2303, S–229.

[26] Interview, Commander George W. Hoover, 5 Mar 1968; "Death of the First Satellite," in Erik Bergaust, *Reaching for the Stars* (Garden City, N.Y.: Doubleday & Co., 1960), p. 214.

[27] See n. 10 and n. 11.

[28] See n. 10 and n. 17; interview, Alan T. Waterman, 15 Oct 1966.

[29] Interview with Milton Rosen, 2 Nov 1966.

Chapter 4

[1] Memo, Deputy Sec/Def for Secretaries of Army, Navy, and Air Force, 9 Sept 1955, Sec/Navy, Cont., S–2303, S–229.

[2] Ltr, Sec/Navy to Ch Naval Research, 27 Sept 1955; ltr, ONR 101:aaj, Ser 001270, 6 Oct 1955.

[3] Rosen diary, 25 Aug 1955.

[4] NRL Conf rpt, 31 August 1955, C–7140–277/55. (Unless otherwise noted, all documents cited in this chapter are in the Vanguard master file in NHF.)

[5] Conf rpts C–7140–273/55, 1 Sept 1955; C–7140–275/55, 7 Sept 1955; C–7140–293, 294, 295, and 309/55, memo rpt, D. G. Mazur, 7–9 Sept 1955. See also John P. Hagen, "The Viking and the Vanguard," in *History of Rocket Technology,* pp. 134–135.

[6] Interviews: Captain Winfred E. Berg, 24 Aug 1966; E. O. Hulburt, 23 April 1965; Daniel G. Mazur, 1 Mar 1967. Kurt Stehling, *Project Vanguard,* pp. 70–71.

[7] See n. 1; handwritten, anon. notes, 13 Sept 1955, among Hagen papers; encl 2 to NRL 4100–227, 10 Oct 1957, and encl 5, "Funding History for Project Vanguard," to NRL 4100–227; interview, Thomas E. Jenkins, 6 Jan 1967.

[8] Conf rpts, 8, 12 Sept 1955; interview, Robert Baumann, 24 Feb 1964.

[9] Conf rpts C–7140–280/55 and C–7140–278/55, 12 and 15 Sept 1955.

[10] Contract Nonr–1817 (00), 23 Sept 1955.

[11] *Ibid.;* Condensed Explanation of Cost Increases for GLM Contract for Vehicle Development, 3 May 1957; interview, Thomas Jenkins, 6 Mar 1967.

[12] Interviews: Capt. Winfred E. Berg, 28 Aug 1966; Milton Rosen, 28 Oct 1966; Elliott Felt and Robert Schlechter, 13 Mar 1967; and Hugh Odishaw, 30 Oct 1967.

[13] Interviews: Elliott Felt, Jr., 13 Mar 1967; Milton Rosen, 6 Dec 1966; Alton Jones, 14 Mar 1967; and John Hagen, 17 Mar 1967.

[14] Rpt on conf, 29 Sept 1955, sub: Vehicle Optimum Weight Distribution. For an account of Martin's use of the analog, see B. Klawans and Joseph E. Burghardt, "The Vanguard Satellite Launching Vehicle, An Engineering Summary," Martin Engineering Report No. 11022, April 1960, p. 7 (hereafter cited as Klawans and Burghardt, Vanguard Launch Vehicle) .

[15] Klawans and Burghardt, Vanguard Launch Vehicle, pp. 7–9, 169; see also Figure 3, "Permissible deviation from nominal path versus range," in Design Specification for Vanguard Launching Vehicle, 29 Feb 1956, revised 29 Mar 1956, Contract Nonr–1817 (00) .

[16] Interviews, Rosen and Bridger, 7 Mar 1968.

[17] Martin purchase order 55–3516–CP, 1 Oct 1955; Martin specification no. 924; Martin Semi-monthly Progress Letters 1 and 2, for period 23 Sept to 23 Oct 1955, pp. 2, 9–10, Martin file for 1956, NHF.

[18] Klawans and Burghardt, Vanguard Launch Vehicle, p. 51.

[19] Rpt on conf, 3 Nov 1955, sub: Air Force contracts with Aerojet-General and Bell Aircraft Corp. to determine past performance; Telephone conf, 9 Nov 1955, sub: Relative performance of Aerojet-General and Bell Aircraft Corp. on Nike liquid booster developments; Martin Semi-monthly Progress Letters 1 and 2, pp. 2, 10–11.

[20] Ltr contract, Martin P.O. 55–3522–CP; Martin specification 925: memo, GLM for BAR, Baltimore, 17 Nov 1955, sub: Prime Contract Nonr 1817 (00) ; interview, Robert Schlechter, 12 Mar 1967.

[21] Rpt, 5 Oct 1955, on conf of 27 Sept 1955; NRL ltr to GLM, 1 Nov 1955; Martin Progress Reports 1 and 2, pp. 11–12; ltr, Schlechter to the authors, 21 Apr 1967.

[22] Martin Vanguard personnel sheets in Martin file, NHF; interview, John Hagen, 17 Mar 1967.

[23] Memo, Dir NRL to GLM, 1 Nov 1955, sub: Project Vanguard Organization; memos, Rosen to Hagen, 9 and 22 Nov 1955, 4140–387/55; Hagen, "The Viking and the Vanguard," in *History of Rocket Technology,* p. 132; Rosen diary.

[24] Ltr, Hagen to C. McL. Green, 8 Jan 1968; ltr, John O'Keefe, Army Map Service, Chief of Research and Analysis, Geodetic Division, to von Braun, 15 Oct 1956, copy in NHF.

[25] Interview, Hagen, Capt. Winfred E. Berg, and Thomas Jenkins, 19 Mar 1968.

[26] Memo, Director NRL to Chief of Naval Research, 30 Sept 1955, and to AF Dpty Chief of Staff for Development, HQ, 2 Nov 1955; 1st Endorsement, HQ, Air Research and Development Command to Director NRL, 2 Dec 1955; conf rpt, 5 Dec 1955, sub: Naval Ordnance Missile Test Facility Review of Pre-

liminary Test Plans for Vanguard Launching Operations; minutes, NRL Project Vanguard staff meeting, 2 Dec 1955.

[27] Memo, Chief of Naval Research to Asst/ Sec Navy (Air), 16 Dec 1955, sub: Scientific Satellite Operation Program: Test Facilities.

[28] Conf rpt, 7 and 9 Dec 1955, to review "Preliminary Test Plans for Vanguard Launching Operations"; memo, Lt. Col. Robert McDaris to MTQX, 14 Dec 1955, sub: Range Safety Considerations for Launching; memo, Asst/Sec/Def Clifford Furnas (R&D) to Asst/ Sec/AF (R&D), 9 Jan 1956, sub: Priority of Earth Satellite Program.

[29] Conf rpt, 20 Sept 1955, sub: Technical requirements for facilities and related technical problems; conf rpt, 12 Sept 1955, sub: Means of gaining GE immediate availability of Malta Test Facility; interviews: Robert Schlechter, 13 Mar 1967, and John Hagen, 17 Mar 1967; ltr, Hagen to C. McL. Green, 22 Mar 1967.

[30] Memo, GLM for Dir NRL, via BAR/Baltimore, 12 Dec 1955, sub: Construction of Vanguard Manufacturing and Testing Installations at GLM; ltr, Ch NR to Asst Sec/Navy (Material), 20 Jan 1956, sub: Facilities Project Associated with GLM Contract, approval of.

[31] Memo for file, 6 Jan 1956, sub: Environmental Test Facility; ltr, GLM to ONR, 4 Jan 1955, sub: Test Installations at GLM.

[32] NRL Memorandum Report (hereafter cited as NRL Memo Rpt) 548, pp. 11, 16–17; memo for the record, 8 Dec 1955, Special Engineering Branch, R&D Division, Troop Operations, Office Ch of Engineers; see also chapter 8.

[33] See chapter 6.

Chapter 5

[1] Hagan notes, to C. McL. Green, 19 Mar 1968; conf rpt, 25 Nov 1955, sub: Facilities for materials research; conf rpt, 9–10 Dec 1955, sub: UDMH heat transfer stability and ignition lags, and high-frequency, lightweight instrumentation. See also Stehling, *Project Vanguard*, pp. 127–34; and Klawans and Burghardt, Vanguard Launch Vehicle, p. 51.

[2] Interview, Hagen, Captain Winfred E. Berg and Thomas Jenkins, 19 Mar 1968; GLM Semi-Monthly Progress Reports 1 and 2, pp. 5 and 7; 3, pp. 5–6; 4, pp. 5–7; 5, p. 1; 6, p. 5. Ltr, NRL to GLM, 1 Nov 1955; conf rpt, 14 Nov 1955; Project Vanguard Rpt to Asst/ Sec/Def (R&D) Policy Council, 12 Dec 1955, NRL Memo Rpt 548, p. 7.

[3] See chapter 4, n. 26; interview Capt. Winfred E. Berg, 10 Apr 1967.

[4] Conf rpt, 27 Oct 1956, sub: Dovap Equipment; conf rpt, 4 Nov 1956, sub: FPS–16 Radar; interview Hagen, Captain Winfred E. Berg, and Thomas Jenkins, 19 Mar 1968.

[5] Memo, Mazur to Rosen, 25 Nov 1955, sub: Telemetering System Requirements; Martin Semi-Monthly Progress Reports 1 and 2, pp. 6, 9, 3, p. 4, 7, p. 6; interview, Robert Schlechter, 13 Mar 1967; minutes, NRL Project Vanguard staff meeting, 2 Dec 1955.

[6] Memo, GLM to Dir NRL, 28 Nov 1955, sub: Recommended Launching Systems Specification.

[7] NRL Memo Rpt 548, pp. 7, 9–10; memo, Dir/NRL to CNO, 29 Nov 1955, sub: Project Vanguard—Frequency Allocation for Minitrack System; Memo, Robert MacKay, Jr., for Mazur, 23 Nov 1955, sub: Relative Qualifications of Bidders . . . Telemetering Ground Stations; conf rpt, 6 Dec 1955, sub: Progress of Antenna Development. See also chapter 9.

[8] Minutes, NRL Project Vanguard staff meeting, 2 Dec 1955; interview Paul Walsh, 18 Apr 1967.

[9] Conf rpt, 6 Dec 1955, sub: NRL–GLM Relationships; conf rpt, 6 Dec 1955, sub: Meeting of Scientific Officer and Operations Manager.

[10] Martin Semi-Monthly Progress Letters 6, p. 5, 10, p. 2; conf rpt, 19 Jan 1956, sub: Weight Optimization Study; NRL Design Specification No. 4100–1, Design Specification for Vanguard Launching Vehicle, Contract Nonr–1817(00), 29 Feb 1956, Appendix 2; interview Paul Walsh, 18 Mar 1967.

[11] Klawans and Burghardt, Vanguard Launch Vehicle, pp. 26–28, 40–51. An example of unexpected problems encountered in preparing the specifications is to be found in Martin Semi-Monthly Progress Letter 13, p. 13. Interviews: Daniel Mazur, 2 Mar 1967; Alton Jones, 14 Mar 1967; and Milton Rosen, 28 Mar 1967.

[12] See NRL Design Specification No. 4100–1, 29 Feb 1956, p. 18, and Appendix 2, p. 1; interview, Rosen, 28 Mar 1967. Like many another decision critically important to the Vanguard program, no written record exists of how this agreement came about. Only the

ensuing action testifies to its validity and only Rosen's memory supplies an oral version of the circumstances.

[13] NRL Design Specification No. 4100-1, pp. 18-20; Stehling, *Project Vanguard*, p. 132.

[14] Martin Semi-Monthly Progress Letters 5, p. 3, 6, p. 3, 9, pp. 5-6; Klawans and Burghardt, Vanguard Launch Vehicle, pp. 51-62; Conf rpt, 19 Jan 1956, sub: Weight Optimization Study.

[15] Stehling Report, 20 Jan 1956, sub: Engine Model Specifications; Comparison of Vanguard Vehicle Delivery and Launch Schedules, no date, but after the autumn of 1957; conf rpt, 9-10 Feb 1956, sub: Agreement on Martin Spec. 925, Propulsion System for Second Stage; interviews, Elliott Felt, 22 June 1967, and Capt. Berg, 10 Apr 1967.

[16] Ltr, GLM to ONR, 19 Jan 1956, sub: Contract Nonr 1817(00), Return of Amendment 1; Ltr, CNR to Dir NRL, 23 Jan 1956, sub: Extension of Period of Interim Contract; memo, Dir NRL for ONR, 31 Jan 1956, sub: Extension of Interim Contract Nonr-1817(00).

[17] NRL Design Specification No. 4100-1, 29 Feb 1956, Revised 29 Mar 1956, Appendix 1, p. 2, and Appendix 2, pp. 1-2. See pp. 79-80.

[18] *Ibid.*, pp. 18-20; Klawans and Burghardt, Vanguard Launch Vehicle, pp. 59, 70-71; interview, Robert Schlechter, 13 Mar 1967; Stehling, *Project Vanguard*, pp. 132-34.

[19] Martin Semi-Monthly Progress Letters 9, p. 5, 11, p. 2; memo, GLM to NRL, 3 Jan 1956; ltr, NRL to GLM, 20 Jan 1956; ltr, GLM to NRL, 3 Feb 1956; ltr, NRL to GLM, 29 Feb 1956; ltr, Van Allen to Hagen, 1 Feb 1956, attachment to minutes 3rd Meeting Technical Panel on Earth Satellite Program, 31 Jan 1956; interviews, Hugh Odishaw, 12 Apr 1967; Paul Walsh, 18 Apr 1967, and John Hagen, 19 Mar 1968.

[20] Conf rpt, 20 Jan 1956, sub: Engine Model Specs; ltr, GLM to BAR, 24 Feb 1956; ltr, GLM to ONR, 19 Jan 1956, sub: Amendment No. 1 to Contract; Martin Progress Ltr 9, pp. 5-6, 10, p. 5, 11, p. 4; conf rpt, 6 Feb 1956, sub: Evaluation of Final Third Stage Rocket Proposals; NRL Design Specification No. 4100-1, 29 Feb 1956, pp. 22-23, and Appendix 2, pp. 1-2.

[21] Ltr, GLM to NRL, 3 Feb 1956, sub: Request for Approval of Clearance of Subcontractors for Third Stage Engine; ltr, NRL to GLM, 7 Mar 1956.

[22] Interview, Milton Rosen, 28 Mar 1967.

[23] NRL Design Specification No. 4100-1, 29 Feb 1956, revised 29 Mar 1956, p. 2, Appendixes 1 and 3; interviews, Elliott Felt, 13 Mar 1967, and Capt. Berg, 10 Apr 1967.

[24] NRL Design Specification No. 4100-1, 29 Feb 1956, pp. 4-6; interview, Milton Rosen, 28 Mar 1967; memo of telephone call, Baumann for Mensshing, 14 Jan 1956; Stehling, *Project Vanguard*, p. 33.

[25] NRL Design Specification No. 4100-1, p. 6; interviews Milton Rosen, 28 Nov 1966, and Paul Walsh, 18 Apr 1967.

[26] Interview, Capt. Berg, 10 Apr 1967.

[27] NRL Design Specification No. 4100-1, pp. 9, 11.

[28] *Ibid.*, pp. 24-30.

[29] Interview, Hagen and Captain Berg, 19 Mar 1968; anon. notes on meeting 13 Sept 1955, cited in chapter 4, n.7. See also chapters 7, 9, and 13.

[30] Memo, GLM to CNR, 8 Mar 1956, sub: Submittal of Proposal for Vanguard Program; NRL Design Specification No. 4100-1, revised 29 Mar 1956, and 4100-1, addendum 1, 27 Apr 1956.

Chapter 6

[1] Minutes of the First Meeting of the Technical Panel on Earth Satellite Program, 20 Oct 1955, and attachment 1 (hereafter cited as min mtg TPESP); National Academy of Sciences—National Research Council, *Report on the U.S. Program for the International Geophysical Year*, IGY General Report, No. 21 (IGY World Data Center A Series), November 1965, p. 553 (hereafter cited as IGY Report 21).

[2] Min 1st mtg TPESP, pp. 3-4, and attachment 2.

[3] Minutes, ad hoc mtg on Earth Satellite Program Budget, 10 Nov 1955; min. 2d mtg TPESP, 21 Nov 1955, and attachments 1a and 1b.

[4] Senate, Committee on Appropriations, *Hearings, Second Supplemental Appropriation Bill, 1956*, 84th Cong., 2d Sess., 21 Mar 1956, p. 209.

[5] Interview, John Hagen, 17 Mar 1967. Min 1st mtg TPESP, 20 Oct 1955, p. 5; 2d mtg, 21 Nov 1955, pp. 2-3; and 4th mtg, 8 Mar 1956, p. 1.

[6] Min 3d mtg TPESP, 28 Jan 1956, p. 2; and 4th mtg, 8 Mar 1956, pp. 3–4.

[7] Senate, Committee on Appropriations, *Hearings, Second Supplemental Appropriation Bill, 1956*, 84th Cong., 2d Sess., 21 Mar 1956, pp. 220–224; House, Committee on Appropriations, Subcommittee on Independent Offices, *ibid.*, 8 Mar 1956, pp. 428, 434, 444–464.

[8] Min 5th mtg TPESP, 20 Apr 1956, pp. 1–2; interview, Kaplan, 12 Feb 1968; Detlev Bronk, speech at the Vanguard tenth anniversary dinner, 16 Mar 1968.

[9] Min 2d mtg TPESP, 21 Nov 1955, attachment 2.

[10] Min 1st mtg USNC Executive Committee, 12 Nov 1954, attachment A.

[11] Min 3d mtg TPESP, 28 Jan 1956 (pp. 2–3, 5–6, and attachments F and G), 4th mtg, 8 Mar 1956 (pp. 4, 7–10), 6th mtg, 8 Jun 1956 (p. 10); note 3 to NRL B–123, June 1957, Vanguard, Master Plan. For Van Allen's request for a cylinder, see chapter 5, pp. 89–90. For further discussion of the 6.4–inch satellite, see chapter 7, p. 127 and 11, p. 197.

[12] Min 4th mtg TPESP, 8 Mar 1956 (attachment A, p. 2), and 5th mtg, 20 Apr 1956 (p. 2); Funding History of Project Vanguard, encl 2 to NRL, 3001–18, NHF; interview, Thomas E. Jenkins, Vanguard comptroller, 22 Aug 1967.

[13] Min 6th mtg TPESP, 7–8 June 1956, pp. 2–3, 13–15, and attachment D; ltr, Kaplan to Waterman, 15 June 1956, in folder 31.1, Procurement of Launching Vehicles.

[14] Ltr, Porter to Kaplan, 17 July 1956; Attachment A to Min 7th mtg TPESP, 5 Sept 1957.

[15] E.g., conf rpt, NRL, 24 Feb 1956.

[16] Cover ltr, Porter to Kaplan, 7 Jan 1957, and encl, "Comments on a Continuing Program of Scientific Research Using Earth Satellite Vehicles."

[17] Interview, Alan T. Waterman, 7 Mar 1967; Funding History of Project Vanguard, NRL, 3001–18; Min 7th mtg, TPESP, 5 Sept 1956, pp. 3–4.

[18] Min 7th mtg TPESP, 5 Sept 1956, pp. 4–6, and attachments B and C; 8th mtg, 15 Oct 1956, pp. 2–4, 8. See also chapter 9, pp. 159–160.

[19] Min 6th mtg, 7 June 1956, pp. 8–9; 7th mtg, 5 Sept 1956, pp. 5–6, and attachment D; 8th mtg, 15 Oct 1956, pp. 2, 4–6; min mtg Working Group on Tracking and Computation, 6 Oct 1956, pp. 1–3.

[20] Min 7th mtg TPESP, 5 Sept 1956, pp. 7–8.

[21] Min 8th mtg TPESP, 15 Oct 1956 (pp. 6–8), and 10th mtg, 7 Feb 1957 (p. 11); "Information about the participation of the Soviet Union in the Investigation of the Upper Atmosphere by means of Rockets and Satellites," encl to memo, Odishaw to the USNC, Executive Committee, TPESP, and Technical Panel for Rocketry, 27 Sept 1956.

[22] Min 8th Mtg TPESP, 15 Oct 1956, pp. 7–8, 9–10, and attachment C.

[23] Min 9th mtg TPESP, 3–4 Dec 1956, pp. 6, 7, 14, 15, and attachments A, B, C, and D; min special session TPESP, 5 Dec 1956, pp. 1–2.

Chapter 7

[1] Cover ltr, James A. Van Allen to Joseph Kaplan, 28 Sept 1955, and enclosure, "A Proposal for Cosmic Ray Observations in Earth Satellites," Satellite Experiments file, folder 32.1, Archives of the National Academy of Sciences: IGY USNC (hereater cited as NAS Archives: IGY USNC).

[2] Interview, James Van Allen, 20 Dec 1967.

[3] "A Proposal for Cosmic Ray Observations in Earth Satellites," encl to ltr, Van Allen to Kaplan, pp. 2–11; min 6th mtg TPESP, 7–8 June 1956, p. 11.

[4] S. Fred Singer, "Measurement of Meteoric Dust Erosion of the Satellite Skin," 20 Oct 1955, Satellite Experiments file, folder 32.2, NAS Archives: IGY USNC.

[5] "Project Vanguard Scientific Research Program," p. 2, attachment 4 to min 2d mtg, TPESP, 21 Nov 1955; interview, Homer Newell, 19 Dec 1967.

[6] Min 2d mtg TPESP, 21 Nov 1955, attachment 4, pp. 2–3.

[7] *Ibid.*, pp. 2, 4.

[8] *Ibid.*, pp. 3–4.

[9] James A. Van Allen, ed., *Scientific Uses of Earth Satellites* (Ann Arbor: Univ. of Michigan Press, May 1956).

[10] Papers presented at the symposium, 26–28 Jan 1956.

[11] Min 1st mtg WGII, 2 Mar 1956, pp. 1–3.

[12] See n. 10 and chapter 9, pp. 155f.; min 3d mtg TPESP, 29 Jan 1956, p. 3, and attachments C, F, and G; Cosmic Ray Observations, proposed by James A. Van Allen, 20 Jan 1956, Satellite Experiments file, folder 32.1, NAS Archives: IGY USNC.

[13] Min 1st mtg WGII, 2 Mar 1956, pp. 4–8.

[14] *Ibid.,* pp. 5–8.

[15] Interview, John W. Townsend, 12 Mar 1968.

[16] Satellite Experiments file, folders 32.1, 32.2, 32.4, 32.6, 32.7, and 32.8, NAS Archives: IGY USNC; min 1st mtg WGII, 2 Mar 1956, pp. 5–7, and 5th mtg 6 Feb 1957; min 6th mtg TPESP, 7 Jun 1956, p. 7.

[17] Folder 32.3, Satellite Experiments file, NAS Archives: IGY USNC; min 2d mtg WGII, 4 Jun 1956, pp. 5–11; min 6th mtg TPESP, 7 Jun 1956, pp. 4, 7. For fuller description of the experiments that received financial support see Lloyd V. Berkner, ed., "Manual on Rockets and Satellites," vol. 6 of *Annals of the International Geophysical Year* (New York: Pergamon Press, 1958) , pp. 310–343; and IGY Report 21, pp. 593–624.

[18] Min 2d mtg WGII, p. 4; min 6th mtg TPESP, 7 June 1956, p. 7.

[19] Min 4th mtg TPESP, p. 6; 5th mtg 20 Apr 1956, p. 3; and 6th mtg, 7 June 1956, pp. 4, 7.

[20] See J. G. Reid, "Status of Internal Experiments for Satellites," 23 Oct 1956, general correspondence file on Internal Instrumentation, 32; Satellite Experiment file, 32.9, NAS Archives: IGY USNC; min 4th mtg WGII, 3 Dec 1956 (p. 4), and 5th mtg, 6 Feb 1957 (p. 17) ; min 9th mtg TPESP, 3–4 Dec 1956, p. 13; Robert W. Stroup, "The Vanguard Satellite Experiments," Naval Research Laboratory, 4 Feb 1958, pp. 7–8.

[21] Min 4th mtg WGII, pp. 7–8; min 9th mtg TPESP, pp. 9–10; Stroup, *op. cit.,* pp. 8–9.

[22] *Ibid.;* Verner E. Suomi, "Radiation Balance of the Earth," Satellite Experiments file, folder 32.11, NAS Archives: IGY USNC; min 4th mtg WGII, 3 Dec 1956, p. 11, and 5th mtg, 6 Feb 1957, pp. 19–21.

[23] Min 5th mtg WGII, pp. 15–21; min 9th mtg TPESP, 3–4 Dec 1956, pp. 9–10, and 10th mtg, 7 Feb 1957, pp. 4–6.

[24] Satellite Experiments file folders, 32.2, 32.3, 32.10, 32.12, NAS Archives: IGY USNC; see also IGY Report, 21, pp. 565, 598, 600, 615, 621; min 11th mtg TPESP, 1 May 1957 (p. 7), and 16th mtg, 12 Feb 1958 (p. 9); Richard Porter notes on draft of this chapter. See also chapters 1, pp. 15–16, and 3, pp. 45–46.

[25] Min 9th mtg TPESP, 5 Dec 1956, p. 12; 10th mtg, 7 Feb 1957, pp. 8–9; 11th mtg, 1 May 1957, p. 9.

[26] Interviews, John Townsend and Robert W. Stroup, 12 Mar 1968.

[27] Interview, Robert C. Baumann, 13 Mar 1968; report on conference at NRL, 24–26 April 1957; attachment A to min 11th mtg TPESP.

[28] Richard Tousey, "Visibility of an Earth Satellite," *Astronautica Acta,* vol. 2, 1956, p. 2; interviews, Tousey, 28 Mar 1968; Whitney Matthews, 14 Mar, and Milton Schach, 27 Mar 1968; report on conference, 24–26 April 1957.

[29] Interviews, Townsend, 12 Mar, and Milton Schach, 27 Mar 1968.

[30] Report on conference, 24–26 Apr 1957; IGY Report 21, pp. 625–626; interviews, John Townsend, Robert Stroup, Whitney Matthews, Robert Baumann, 12 Mar 1968, John Hagen, 19 Mar 1968, Milton Schach, 27 Mar 1968, and Richard Tousey, 28 Mar 1968.

[31] Report on conference, 24–26 April 1957; IGY Report 21, pp. 625–626; and min 11th mtg TPESP, p. 8, and attachment B, pp. 2–3.

[32] Ltr, Hagen to Porter, 4 Sept 1956, attachment G to min 7th mtg TPESP, Sept 1956; conf rpt, 24–26 Apr 1957.

[33] Private communication to the author.

[34] Funding History of Project Vanguard, encl 2 to NRL 3001–18, NHF.

[35] Interview, Thomas E. Jenkins, 6 Mar 1967.

[36] Senate, Committee on Appropriations, *Hearings on Supplemental Appropriation Bill for 1958,* 85th Cong., 1st sess., 2 Aug 1957, p. 98.

[37] *Ibid.,* 98–99, 110–117.

[38] Interview, Milton Rosen, 17 Nov 1966.

[39] See chapter 10.

[40] Interview, Odishaw, 4 Oct 1967; min 12th mtg TPESP, 3 Oct 1957, pp. 3–4.

Chapter 8

[1] Eugene M. Emme, *Aeronautics and Astronautics: An American Chronology of Science and Technology in the Exploration of Space, 1915–1960* (Washington, D.C.: NASA, 1961) , pp. 62, 65, 67 (cited hereafter as NASA Chronology). See also Stanley J. Macko, *Satellite Tracking* (New York: J. F. Rider, 1962) , chapter 10.

[2] Memo, C–7140–309/55, 7 Sept 1955 (AFMTC notified unofficially that they have been chosen for launching site); memo, Code 7140 to Code 1900, 15 Sept 1955, sub: Project Vanguard; information for stub requisition

for contract with GLM, C–7140–278/55; ltr, Dir/NRL to Deputy Chief Staff (Develop), USAF, 2 Nov 1955, sub: Scientific Satellite Operation Program (Project Vanguard) ; ltr, Dir/NRL to Chief/Naval Research, 28 Nov 1955, sub: Use of PAFB for Project Vanguard, C–4106–5/55.

³ Ltr, Hq ARDC to Dir/NRL, 2 Dec 1955, sub: Scientific Satellite Operation Program (Project Vanguard) .

⁴ Memo, Sec/Def to Secs/Army, AF, and Navy, Sec/Nav Cont. No. S–2303 S–229, 9 Sept 1955, sub: Technical Program for NSC 5520 (capability to launch a small scientific satellite during IGY) ; min staff mtg, Project Vanguard, 2 Dec 1955.

⁵ Min staff mtg, Project Vanguard, 2 Dec 1955.

⁶ Interview, Robert Schlechter of Martin, 13 Mar 1967.

⁷ The consultations at Patrick in Sept 1955 are covered in the following Consultative Service Records (conference reports) in NHF: Ser. C–7140–292/55, –293, –294, –295, and –296 of 15 Sept; and –305, –308, –309, –310, –311, –312, and –315 of 29 Sept. Except as otherwise indicated further references to the September talks at the Cape rest on these records.

⁸ Interview, 2 Mar 1967.

⁹ USAF memo of 23 Sept 1955.

¹⁰ Stehling, *Project Vanguard*, p. 93; interview with Sears William of Martin, 5 May 1967.

¹¹ Undated speech, "Vanguard Operations at Atlantic Missile Range" (one of the designations of the Cape Canaveral test center) .

¹² Interview, Captain Winfred E. Berg, USN, 19 Apr 1967. See also minutes of Technical Board meeting 6–57, dated 5 Feb 1957, and ltr 4106–11 Ser: 13567, Dir/NRL to Deputy Chief of Staff/Development, Hq. USAF, 19 Nov 1957, sub: Project Vanguard, Air Force Program Offers for.

¹³ Mazur, interview.

¹⁴ Ltr, Dir/NRL to Hq. AFMTC, 16 Nov 1955, sub: Vanguard Preliminary Plan, Transmittal of; ltr, Gibbs to Hq. Research and Development Command, 28 Nov 1955, sub: Preliminary Planning for Project Vanguard; conf rpt, 5 Apr 1956, 4120–152/56; "Case History" by Cdr. Calhoun, Apr 1956, PAFB 56–8914; ltr, Dir/NRL to Hq. AFMTC, 2 May 1956, sub: Vanguard Test Program, Transmittal of; Project Directive No. 2, 11 May 1956, sub: Vanguard Test Program Changes; interview, Schlechter.

¹⁵ Berg interview. See also Stehling, *Project Vanguard*, p. 93.

¹⁶ Memo, Asst Sec/Def for Sec/Navy, 9 Jan 1956, sub: Scientific Satellite Operation Program (Project Vanguard; Test Facilities) ; Trevor Gardner Memo to Asst Sec/Navy (Air) , 10 Jan 1956, sub: Air Force Approval of AFMTC for Vanguard Use; Berg interview; "Diary of Project Vanguard" by Joseph E. Burghardt, Martin's assistant project engineer for aerodynamics and propulsion and later technical director for systems, 2 Feb 1956. This valuable personal record kindly lent to the authors by Mr. Burghardt is cited hereafter as Burghardt Ms. Diary.

¹⁷ Wernher von Braun, "The Redstone, Jupiter, and Juno," *History of Rocket Technology*, pp. 107–111; Robert L. Perry, "The Atlas, Thor, Titan, and Minuteman," *ibid.*, pp. 150–151; Martin Semi-Monthly Progress Letter 9, p. 7; min staff mtg, Project Vanguard, 3 Jan 1956; conf rpt 4150–6/56 eb of 23 Jan 1956; minutes of conference at Redstone Missile Firing Branch, PAFB, 26 Jan 1956; conf rpt S4130–2/56, 17 Jan 1956; GLM inter-department communication of 6 Feb 1956, Schlechter to Markarian, sub: Facilities for Vanguard Field Program; conf rpt 4110–140/56 of 8 Jan 1956 and attached minutes; Klawans and Burghardt, Vanguard Launch Vehicle, pp. 99–101; Project Vanguard Report 2 (NRL Report 4717), p. 3; and Schlechter interview.

¹⁸ Ltr, Dir/NRL to Chief of Engineers for Military Construction, 16 Feb 1956, sub: Transfer of Gantry Crane from White Sands . . . to AFMTC, Ser: 3601; ltr, 0881, GLM to Dir/NRL, 17 Feb 1956, sub: Viking Spare Parts and Equipment at White Sands Proving Ground, Evaluation of Present Usability; ltr, SAKML–2 (AFMTC) , Corps of Engineers to Dir/NRL, 20 Apr 1956, sub: Relocation of Gantry Crane, Patrick Aux. No 1 Air Force Base; Cdr. Calhoun Newsletter 2–56715, Mar 1956; contract Da–08–123–ENG–2049, 24 Apr 1956, Construction of Launching Complexes Nos. 17 and 18 at AFMTC, in Mail and Records Section, Canaveral District, Corps of Engineers, Merritt Island, Fla. (cited hereafter as Jones Contract).

¹⁹ Conf rpt 4120–125/56 (4124) , 26 Mar 1956; Vanguard Progress Report 6, p. 1.

²⁰ Ltr 1246 to NRL.

[21] Interview, Mengel, 7 Mar 1967.

[22] Vanguard Progress Report 2 (NRL Report 4717), p. 3, and NRL Report 4800, p. 1; ltr, Chief/Air Force Projects Division, Military Construction, to Division Engineer, South Atlantic Division, Atlanta; Ga., 29 Mar 1956, sub: Authorization-Patrick Aux. No. 1 Air Force Base—Construction of Navy (Vanguard) Facility, Assembly Bldg "S"; memo, Chief/ Naval Research to Chief/Naval Operations, 11 May 1956, sub: Project Vanguard Weekly Status Rpt. No. 4.

[23] NRL Report 4717, p. 4; Jones Contract; conf rpt C4150–6/56, 13 Jan 1956; Schlechter and Mazur interviews.

[24] Berg interview; Martin Semi-Monthly Progress Letter 9, p. 7.

[25] GLM inter-department communication, 2 Feb 1956, sub: Facilities for Vanguard Field Program—PAFB Trips on Jan 26, 27, and Feb 1, 2.

[26] Berg interview; Vanguard Progress Reports 10, p. 1, and 11, p. 3; ltr, Chief/Naval Research to Chief/Naval Operations, 3 Feb 1956, sub: Project Vanguard Weekly Status Rpt.

[27] Ltr, Office/Naval Research to GLM, 31 July 1956, sub: Project Vanguard; Vanguard memo 4104–2/56, Master Planning Coordinator to Project Director, 17 July 1956, sub: Vanguard—Amendment to July Rpts on Significant Items Which are Behind Schedule; ONR memo, ONR:100:dr Ser 01152, 13 Aug 1956, sub: Project Vanguard Weekly Status Rpt. No. 26.

[28] Preliminary "Design Criteria for Static and Flight Firing Structure," dated 21 Feb 1956, in Martin Marietta Corporation files, p. 1; GLM purchase order 56–3529–CPFF, 20 Jan 1957, to Loewy Hydropress Division of Baldwin-Lima-Hamilton Corporation, New York City.

[29] Interviews, 21 Apr 1967, with Berg, Manning, and W. D. Skilochenko of NRL; Martin Monthly Progress Letter 4, p. 8; Jones Contract; case history, Project Vanguard, PAFB 56–12964, July 1956; ibid., PAFB 56–18360, Oct 1956; ibid., PAFB 56–18898, 5 Nov 1956; conf rpt 5230–373/56, 8 Oct 1956; and status rpt, 18 Oct 1956, sub: Vanguard Operations and Facilities at PAFB.

[30] For a discussion of the National Academy's desire for twelve rather than six mission vehicles, see chapter 6, p. 98.

[31] NRL Design Specification No. 4100–1, p.

29; interview, Markarian, 9 Oct 1967; ltr, Chief/Naval Research to GLM, 15 July 1957, sub: Change Order No. 68; memo, Chief/Naval Operations, 23 July 1957, sub: Project Vanguard Weekly Status Rpt No. 75.

[32] PAFB case histories, Project Vanguard, July 1956 and Oct 1956; conf rpt 5230–373/56, 8 Oct 1956; status rpt, 18 Oct 1956, sub: Vanguard Operations and Facilities at PAFB.

Chapter 9

[1] For this and other details connected with the early development of radio tracking, see Corliss, The Evolution of STADAN, chapter 2.

[2] Appendix C of the Naval Laboratory's proposal to the Stewart Committee. Full reference: "A Scientific Satellite Program," NRL Memo Report 487, 5 July 1955, with addendum letter S–7140–262/55, 23 Aug 1955. See also chapter 1, pp. 10–11.

[3] See Whipple, "Photographic Meteor Studies. I," Proceedings of the American Philosophical Society, vol. 79, 1938, pp. 499–548; and his "Meteoritic Phenomena and Meteorites," in Clayton S. White and Otis O. Benson, Jr., eds., Physics and Medicine of the Upper Atmosphere: A Study of the Aeropause (Albuquerque: University of New Mexico Press, 1952), pp. 137–170.

[4] Corliss, op. cit., p. 14.

[5] Ltr, Homer E. Newell, Jr., to Clyde W. Tombaugh, New Mexico College of Agriculture and Mechanical Arts (NMCAM), 20 June 1956; and undated and untitled typescript identifiable by the handwritten insert, "save for minor changes, this is it," in Hagen's personal file folder.

[6] Encl (1) of ltr 4100–251/56, Dir/NRL to Chief/Information, Navy Dept., 22 Aug 1956, sub: Press Release and Photographs on Minitrack Transmitter.

[7] John T. Mengel, "Tracking the Earth Satellite, and Data Transmission by Radio," Proceedings of the IRE, vol. 44, no. 6 (June 1956), p. 755. See also ltr 2030–37 Ser: 1927, Vanguard Technical Officer to David O. Woodbury, Santa Barbara, Cal., 26 Feb 1957.

[8] See n. 6 and DoD news release 969–56, 17 Sept 1956, sub: Earth Satellite Instruments to be Displayed in New York.

[9] Mengel, op. cit.

[10] John T. Mengel and Paul Herget, "Track-

ing Satellites by Radio," *Scientific American*, vol. 198, no. 1 (January 1958), pp. 23–29.

[11] Attachment C, min 7th mtg TPESP, 5 Sept 1956.

[12] See chapter 3.

[13] NRL Memo Report 487, p. 22; ltr, Hagen to Hugh Odishaw, 9 Aug 1956; NRL Report 4747, p. 16; Easton's ms., "Radio Tracking of the Earth Satellite," 12 Apr 1956, pp. 3–10, and accompanying drawings; encl (1) of ltr, Dir/NRL to Chief/Information, Navy Dept., 13 June 1956, sub: Radio Tracking of Earth Satellite, News Release on; IGY Report 21, Nov 1965, pp. 587f; min 4th mtg TPESP Working Group on Tracking and Computation, p. 3, and min 8th mtg TPESP, p. 8.

[14] Mengel interview; ltr, Chief/Engineers to Dir/NRL, 10 May 1957, sub: Project Vanguard-Army Prime Minitrack Program, a summary of progress; Memo 105 Ser: 9147, Chief/Naval Research to Chief/Naval Operations, 5 Sept 1957, sub: Project Vanguard Weekly Status Rpt No. 81; attachment 1, memo, NRL to James O. Spriggs, Office/Asst Sec. Def, 17 Sept 1957, sub: Draft of Proposed Memo to the President, submittal of.

[15] Min 4th and 5th mtgs TPESP and attachment 10 to min 7th mtg.

[16] IGY Report 21, p. 569; interview, Whipple, 9 Oct 1967.

[17] E. Nelson Hayes, The "Smithsonian's Satellite-Tracking Program . . . [Part 1]," *Annual Report . . . Smithsonian Institution . . . 1961*, pp. 286, 298.

[18] IGY Report 21, p. 568.

[19] Min 12th mtg TPESP.

[20] IGY Report 21, p. 569; Porter to authors, 21 Mar 1968.

[21] Ltr, Chief/Naval Research to Chief/R&D, Office/Chief of Staff, Dept/Army, 10 May 1956, sub: Project Vanguard, Passive Tracking Program For; interviews, Hagen, 17 Mar 1967, and Whipple, 9 Oct 1967.

[22] IGY Report 21, pp. 571f; min 18th mtg TPESP; Hayes, *op. cit.*, p. 289; conf rpt 7340-687, 12 July 1957, third meeting to review and coordinate the joint efforts of NRL, USAF, and SAO on the Moonwatch; ltr, Whipple to Hagen, 20 Dec 1957.

[23] Min 15th and 17th mtgs TPESP.

[24] Minutes of staff meeting, Project Vanguard, 9 Dec 1955.

[25] Ltr 4130-119 Ser: 2374, Dir/NRL to CO/US AMS, 13 Mar 1956, sub: Calibration Aircraft for Minitrack Stations; memo 6260-349/55, NRL 12 Dec 1955, sub: Aircraft Requirements for Equipment Evaluation, Project Vanguard; master ltr, Mengel to Box 983W c/o IRE, 19 Feb 1957.

[26] W. E. Smitherman, "Army Participation in Project Vanguard," *IRE Transactions on Military Electronics*, vol. MIL-4, nos. 2-3 (April–July, 1960), p. 326.

[27] Mengel and Herget, *op. cit.* (see no. 10), p. 27.

[28] *Ibid.*

[29] See chapter 8, p. 141.

[30] NRL ltr 3000-122/56, 15 Aug 1956, Dir/NRL to Asst Sec/Def (Comptroller), sub: Project Vanguard Tracking Stations, Request for Release of Funds to Department of the Army for Construction of; encl (1) NRL ltr 4130-84 Ser: 2168, 7 Mar 1957; NRL memo 4102-71, 18 Mar 1957, sub: Vanguard Communications; CNO ltr Op-445G Ser: 699P44, 2 Mar 1956, Fourth Endorsement on NRL Washington ltr 2501A-295/56 (1), 10 Feb 1956, Chief/Naval Operations to Asst Sec/Navy (Materiel), sub: Project Vanguard; Proposed Minitrack Site Blossom Point Proving Ground; Dept Navy/BuYds & Docks ltr R-421, 12 Apr 1956, Chief/BuYds & Docks to Real Estate Division, Corps Engineers/Depart Army, sub: Diamond Ordnance Fuse Laboratory, Blossom Point Proving Ground . . . Request for Permit to Use Portion of Land; NRL ltr 4130-291/56 (4130), 3 Apr 1956, Dir/NRL to CO/AFMTC, sub: Project Vanguard: Proposed Minitrack Test Plan for; ONR ltr 108 Ser: 27291, 21 Nov 1956, Chief/Naval Research Asst Ch/Staff, Installations, USAF, sub: Vanguard Minitrack Facilities on Downrange Islands; Office Chief/Engineers ltr, 18 Feb. 1957, sub: Project Vanguard—Army Prime Minitrack Program; NRL ltr 2050-90 Ser: 2741, 18 Mar 1957, Dir/NRL to Chief/Information, Navy Dept, sub: "Minitrack," A Chapter of a New Book, "Little New Moon," by David O. Woodbury; Australian Joint Service Staff ltr 613/R64, 2 Apr 1957, to NRL; Corliss, The Evolution of STADAN, chapter 2 and Appendix A; conf rpt 4130-60, 8 Feb 1957.

[31] NRL ltr C-4108-7/55, 12 Jan 1956, Dir/NRL to Chief, Engineering Intelligence Division, Office Chief/Engineers, sub: Study for Tracking Sites for Project Vanguard; encl (1) to NRL ltr C-4106-16/56 of 12 Mar 1956, sub: Preliminary Specifications and Considerations, sub: Minitrack Site Facilities; NRL Report

4767, pp. 1f; Smitherman, *op. cit.* (see n. 26) ; NEL memo, 24 May 1956, sub: Project Vanguard, NEL Assistance in; NRL ltr C4100–128/56, 5 June 1956, Dir/NRL to Chief, Bureau of Ships, sub: Authorization for NEL Participation in Project Vanguard, Request for; interviews, 19 June 1967, with James J. Fleming and Joseph W. Siry.

[32] "Specification for Minitrack Ground Station Units," 28 May 1956; NRL ltr C–4130–57/56 (4130) 100317, 19 July 1956, Dir/NRL to Chief/Naval Operations (Contracts Div.), sub: Minitrack Ground Station Units, Selection of Contractor for; Contract Nonr–2190 (00), 1 Aug 1956, between Office of Naval Research and Bendix Radio Div. of Bendix Aviation

[33] NRL ltr 4130–34/56 (4130) of 9 May 1956, Corp., Baltimore, Md. Dir/NRL to CO/AFMTC, sub: Project Vanguard: Minitrack Antenna Mounts; CSR 5330–500/56, 13 June 1956; CSR 5330–574/56, July 1956; ONR ltr 610 Ser: 17999, 25 Jan 1957, Chief/Naval Operations to Technical Appliance Corp., sub: Contract Nonr–2268 (00) .

[34] "Melpar, Inc." contract file; encl (1) of NRL ltr 4100–259/56, 7 Sept 1956, Dir/NRL to Chief/Information, Navy Dept, sub: AN/DPN (48) Radar Beacon (article for Melpar Company publication) .

[35] ONR ltr 601 Ser: 17854, 6 Aug 1956, Bennett to Dir/NRL.

[36] Interviews: Herget, by phone, 18 July 1967; Mengel, 17 Mar 1967; Siry and Fleming, 19 June 1967.

[37] NRL ltr 4100–161/56 (110044), 26 June 1956, Dir/NRL to Supt, U.S. Naval Observatory, sub: Assistant in Project Vanguard, Request for; Dept/Navy, U.S. Naval Observatory, ltr Ser: 1078, 29 June 1956, Supt/Naval Observatory to Dir/NRL; Encl (1) to NRL ltr 4100–189/56, 17 July 1956, sub: Vanguard Project Directive, Working Group on Orbits, Establishment of.

[38] NRL ltr C–7100–149/55, 14 Oct 1955, Dir/NRL to Dr. C. C. Hurd, IBM, 590 Madison Avenue, New York City 22, sub: Computer Needs for Project Vanguard.

[39] *Ibid.*

[40] ONR ltr 610J, 9 Mar 1956; NRL ltr C–4100–112/56, 8 May 1956, Dir/NRL to Chief/Naval Research, sub: Contractor for Satellite Orbit Computational Facilities for Project Vanguard, Selection of; Nonr–2169 (00) , Contract between Government and IBM, dated 22 June 1956; NRL news release, 5 July 1956, sub: IBM to Provide Computing Facility for Project Vanguard; IBM ltr, 15 Apr 1957, sub: Project Vanguard Progress Rpt 9; news release (source not indicated) , slugged "For Release: June 30, 1957 at 8 p.m.," and dated Washington, June 30, 1957; IBM ltr, 30 Aug 1957, to NRL, sub: Project Vanguard Progress Rpt 13.

[41] NRL Report 5113, p. 17; memo, Hagen to Codes 1510 and 1810, 4 June 1956, sub: Dr. Peter Musen, Information Covering; and A. Robin Mowlem, mathematician in charge of programming, IBM Computing Center, "How the Computer Calculates Orbits," [undated IBM information release].

[42] Mazur, Fleming, and Siry interviews; NRL memo 4150–85/56 (4150) , 6 June 1956, sub: Justification for Use of AN/FPS–16; minutes of Technical Board meeting 6–57, 5 Feb 1957; memo 5130–17: ABB, 1 Mar 1957, sub: Conference Rpt, Data Transmission Link between IBM 704/sic/Computer and Third-Stage Firing Console . . . ; CSR 5230–158, 10 Apr 1957, sub: "Ground Backup Guidance Plan Discussed"; NRL Report 5113, p. 17.

[43] NRL Technical Information Division release, 9 July 1957, "Navy Will Use 'Electric Brain' to Study Flight of Vanguard Launching Vehicles"; Mengel and Herget, *op. cit.* (see n. 10) , pp. 27f; NRL Report 4880, p. 23; NRL Report 5113, p. 17; Fleming-Siry interview.

Chapter 10

[1] Interview, 13 Mar 1967.

[2] Encl (1) to ltr, Dir/NRL to GLM, 15 Sept 1956, sub: Organization at AFMTC Vanguard Project directive, 14 Sept 1956, sub: Organization & Mode of Operation at AFMTC; ltr 4100–226, Dir/NRL to Maj. Gen. Donald N. Yates, AFMTC, 18 Sept 1956; ltr, U.S. Naval Unit Hq., AFMTC, to Chief/Naval Operations, 1 Jun 1956, sub: Proposed U.S. Naval Facility at AFMTC, Status of; ltr 1914, GLM to Dir/NRL, 10 Aug 1956, sub: Assignment of Personnel to Vanguard Field Crew; ltr 4100–16t, Ser: 8815, Dir/NRL to Yates, 9 Aug 1957; memo 4100–28, Code 4100 (Hagen) to Records, 3 Feb 1958, sub: Relations Among Project Vanguard Personnel at AFMTC; ltr, Schlechter to authors, 21 Apr 1967; interview, Gray, 23 May 1967.

[3] Schlechter interview and ltr; Stehling, *Project Vanguard*, p. 78.

[4] Interview, John R. Zeman, 22 and 23 Mar 1967.

[5] Zeman interview; ltr, code 4160 (Mazur) to Code 1000, 1 July 1957, sub: Prolonged Temporary Duty, Per Diem, Justification for; memo, Hagen to Code 1000, 24 Apr 1957, sub: Change in Effective Date of Reduced Per Diem, 4100–104.

[6] Zeman interview; Stehling, *Project Vanguard*, pp. 173f; memo, Chief/Naval Research to Chief/Naval Operations, 29 Jan 1958, sub: Project Vanguard Weekly Status Rpt No. 97.

[7] Stehling, *Project Vanguard*, p. 194.

[8] Zeman and Mazur interviews.

[9] Klawans and Burghardt, "The Vanguard Launching Vehicle: An Engineering Summary," Martin Engineering Report No. 11022, April 1960, p. 98; conf rpt 4120–85, 29 Jan 1957, Purpose: To establish acceptance procedural policy for Vanguard vehicle acceptance and delivery; interviews: Felt, 13 Mar 1967, and Bridger, 3 Aug 1967.

[10] Interview, 3 Aug 1967; ltr 4100–19, Dir/NRL to GLM, 1 Nov 1955, sub: Project Vanguard organization; memo 4100–55, Code 4100 to Distribution List, 5 Dec 1955, sub: Organization of Technical Director's Staff. See also Rosen, *The Viking Rocket Story*, chapter 10.

[11] Klawans, and Burghardt, *op. cit.*, pp. 104–06; see also encl (4), ltr, Dir/NRL to Chief/Naval Research, 19 June 1958, sub: Contract Nonr 1817 (00), Project Vanguard, Proposed SCN Nos. 1, 2, and 3 to NRL Specification No. 4100–1, Submittal of.

[12] But see chapter 12, p. 218, for change in countdown procedure from TV–4 on.

[13] Klawans and Burghardt, *op. cit.*, pp. 109f; [Vanguard directive] 4130–200, "Installation Procedure for Minimal Satellite;" encl (1) to ltr 4104–33, Planning Coordinator to VOG Field Manager, sub: Draft of OR–702 for TV–3; interviews, Bridger and Gray. See on scientific considerations *vis à vis* launch times, encl (1) to NRL ltr 4100–74, 1 Aug 1956, memo for file, sub: Feasibility Study for a Five-Week Launching Cycle for Vanguard Vehicles.

[14] Klawans and Burghardt, *op. cit.*, p. 112.

[15] Interview with Schlechter; NRL Rpt 5183, p. 14; NRL Rpt 5185, p. 6; memo, Ser: 01722, Chief/Naval Research to Chief/Naval Operations, 18 Dec 1956, sub: Project Vanguard Weekly Status Rpt 44; encl (1) OSD memo, 15 Oct 1957, sub: Rough draft of Project Vanguard Rpt for Submission to Senate Armed Services Committee; Klawans and Burghardt, *op. cit.*, p. 112; encl (1) Case History, Project Vanguard, December 1956, sub: Post-Launch Briefing on Vanguard Test Vehicle TV–0.

[16] Encl (1), Case History, Project Vanguard, Dec 1956, sub: Post-launch briefing of Vanguard Test Vehicle TV–0; Burghardt Ms. Diary, *passim;* Klawans and Burghardt, *op. cit.*, p. 112; ltr 1851, GLM to Dir/NRL, 23 July 1957, sub: Master Planning Chart, Submittal of.

[17] See chapter 8, pp. 142–143.

[18] Memo, Chief/Naval Research to Chief/Naval Operations, 26 Dec 1956, sub: Project Vanguard Weekly Status Rpt No. 45; ltr 0661, GLM to Chief/Naval Research, 7 Mar 1957, sub: Project Vanguard Changes Requested for Incorporation at AFMTC; memo, Chief/NRL to Chief/Naval Operations, 16 Jan 1957, sub: Progress Weekly Status Rpt No. 42; ltr, GLM to Dir/NRL, 25 Jan 1957, sub: Delivery of Test Vehicle No. 1; encl (1) to ltr, Dir/NRL to Chief/Naval Operations, 15 Feb 1957, sub: Vanguard Requirements for AFMTC, 4106–6, Ser: 9238; memo, Rosen to Hagen, 19 Feb 1957, sub: Case Histories of Problem Areas, 4110–12; memo, Chief/Naval Research to Chief/Naval Operations, 15 Mar 1957, sub: Project Vanguard Weekly Status Rpt No. 59; encl (1) to ltr 4100–89, Dir/NRL to Chief/Naval Research, 5 Apr 1957, sub: Project Vanguard Weekly Status Rpt No. 62; Project Vanguard Master Planning Coordinator, Monthly Progress Report of Significant Items, 4104–18, 1 May 1957; memo, Chief/Naval Research to Chief/Naval Operations, 7 May 1957, sub: Project Vanguard Weekly Status Rpt No. 64; Klawans and Burghardt, *op. cit.*, p. 112; Schlechter interview.

[19] Mazur interview.

[20] Klawans and Burghardt, *op. cit.*, p. 112.

[21] Interviews, Bridger and Donald J. Markarian on 12 Oct 1967; see also Klawans and Burghardt, *op. cit.*, p. 40.

[22] Ltr 4100–132, Ser: 0878, Dir/NRL to GLM, 10 June 1957, sub: Provisional Acceptance of TV–2; and conf rpt 4110–28, 13 June 1957, purpose: Provisional Acceptance of TV–2, Discussion of.

[23] Ltr 1875, GLM to Chief/Naval Research, 30 July 57, sub: Modifications of Vehicles and Equipment After Delivery, Provisions for.

[24] Encl (1) to ltr 01077, Dir/NRL to Chief/

Naval Research, 12 July 1957, sub: Project Vanguard Weekly Status Rpt No. 76.

²⁵ Project Vanguard master planning coordinator, Monthly Progress Report, 1 June 1957, 4104–23; Klawans and Burghardt, *op. cit.*, p. 102; GLM inter-department communications of 19 June, 8 July, and 10 July, 1958, Martin-Marietta Corp., archives; interview, Sears Williams, 13 Mar 1967.

²⁶ GLM Monthly Progress Letter 17, p. 8; Weekly Progress Report 4104–44, 26 Aug 1957, Vanguard Planning Office to Vanguard Project Director.

²⁷ Mazur and Schlechter interviews.

²⁸ Hagen and Rosen interviews.

²⁹ Ltr 4103–23 Ser:11206, Dir/NRL to GLM, 21 Oct 1957, sub: Static Firing of TV–2, Requirement for.

³⁰ Private information; and see Stehling, *Project Vanguard*, p. 85, and R. C. Hall, "Vanguard and Orbiter," *The Airpower Historian*, vol. 9, no. 4 (October 1964), pp. 109f.

³¹ GLM Monthly Progress Letter 17, p. 9; encl (1) to ltr 4100–191 Ser:91322, Chief/Naval Research to Chief/Naval Operations, 30 Aug 1957, sub: Project Vanguard Weekly Status Rpt No. 83; Burghardt Ms. Diary, 8–28–57; Weekly Progress Report 4014–47, Vanguard Planning Office to Vanguard Project Director, 3 Sept 1957; encl to memo for James O. Spriggs, Office Asst Sec/Defense, 17 Sept 1957, sub: Draft of Proposed Rpt to the President, Submittal of; memo, Chief/Naval Research to Chief/Naval Operations, 12 Sept 1957, sub: Project Vanguard Weekly Status Rpt No. 82; ltr GE to Dir/NRL, 17 Sept 1957, sub: Return of X–405 Engine No. SN–9.

³² GLM Monthly Progress Letter 17, p. 9; encl (1) to ltr 4100–107 Ser: 0142, Dir/NRL to Chief/Naval Research, sub: Project Vanguard Weekly Status Rpt No. 85; GLM Monthly Progress Letter 17, pp. 9, 10; *ibid.*, 18, p. 14; ltr 2303, GLM to Dir/NRL, 12 Sept 1957, sub: Project Vanguard Static Firing of TV–2, Requirement for: conf rpt 4110–42, 20 Sept 1957, purpose: To Discuss Plan of Operation of TV–2; ltr 2503, encl (1) to ltr 4100–213 Ser: 01446, Dir/NRL to Chief/Naval Research, 23 Sept 1957, sub: Project Vanguard Weekly Status Rpt. No. 86; encl (1) to ltr 4100–214 Ser: 01491, Dir/NRL to Chief/Naval Research, 30 Sept 1957, sub: Project Vanguard Weekly Status Rpt. No. 87; Weekly Progress Report 4104–60, Vanguard Planning Office to Van-

guard Project Director, 8 Oct 1957; encl (1) to ltr 4100–299 Ser: 01562, Dir/NRL to Chief/Naval Research, 14 Oct 1957, sub: Project Vanguard Weekly Status Rpt. No 88; encl (1) to ltr 4100–271 Ser: 01651, Dir/NRL to Chief/Naval Operations, 22 Oct 1957, sub: Project Vanguard Weekly Status Rpt. No. 89; Klawans, and Burghardt, *op. cit.*, p. 112.

³³ Stehling, *Project Vanguard*, p. 122.

Chapter 11

¹ *New York Times* (cited hereafter as NYT), 1 and 2 Oct 1957; E. Nelson Hayes, "The Smithsonian's Satellite-Tracking Program: Its History and Organization, Part 2," *Annual report . . . Smithsonian Institution . . . 1963*, pp. 334–336 (cited hereafter as Smithsonian Tracking Program:2); Hayes, "Tracking Sputnik I," in Arthur C. Clarke, ed., *The Coming of the Space Age* (New York: Meredith Press, 1967), p. 10; U.S. Senate, Committee on The Armed Services, Preparedness Investigating Subcommittee, *Hearings: Inquiry into Satellite and Missile Programs*, part 1, 85th Cong., 1st and 2d Sess., p. 156 (cited hereafter as Johnson Subcommittee Hearings, 1957).

² NYT, 31 Aug, 1 Sept, 1 and 5 Oct 1957; Walter Sullivan, *Assault on the Unknown: The International Geophysical Year* (New York: McGraw-Hill Book Co., 1961), pp. 1f; phone interview with Sullivan; 21 Mar 1968; Richard Porter to authors, 21 Mar 1968.

³ NYT, 12 Oct 1957; NASA Chronology, p. 91 and appendix A; Hayes, Smithsonian Tracking Program:2, pp. 336f.

⁴ Dwight D. Eisenhower, *The White House Years; Waging Peace, 1956–1961* (New York: Doubleday & Company, Inc., 1965), p. 206 n.

⁵ NYT, 5 Oct, 5 and 12 Nov, and 5 Dec 1957; Richard Witkin, ed., *The Challenge of the Sputniks, in the Words of President Eisenhower and Others* (Garden City, N.Y.: Doubleday & Co., 1958), pp. 3, 4, 18; see also memo for Sec. Nav . . . , 9 Oct 1957, sub: Briefing of Preparedness Investigating Subcommittee Staff on the Vanguard Program and the Significance of the Russian Satellite Launching.

⁶ Witkin, *op. cit.*, pp. 6, 34–50; NYT, 5, 6, and 10 Oct, 8, 12, and 14 Nov 1957; Eisenhower, *op. cit.*, p. 211.

⁷ Witkin, *op. cit.*, pp. 12, 13, and 16; NYT, 6, 12, 13, 14, and 15 Oct 1957; Hayes, Smithsonian Tracking Program:2, p. 334; Whipple

interview; see also Johnson Subcommittee Hearings 1957, pt. 1, pp. 45f and 165. See also Sullivan, *op. cit.* (see n. 2), pp. 64, 70.

[8] NYT, 5 Oct 1957; Witkin, *op. cit.*, p. 6.

[9] W. E. Smitherman, "Army Participation in Project Vanguard," *IRE Transactions on Military Electronics*, vol. MIL–4, Nos. 2–3 (April–July, 1960), p. 326; interview, Siry; min 11th mtg TPESP, 1 May 1957.

[10] Hayes, Smithsonian Tracking Program:2, p. 335.

[11] Ltr, IAGS Peru Project to Dir/Project Vanguard Attn Code 4130 [Mengel] 14 Oct 1957, sub: Weekly Project Report, Lima Station, Week ending 12 Oct 57.

[12] Smitherman, *op. cit.*, p. 326; Progress Report, Project Vanguard, 9 Oct 1957; NYT, 1 Oct 1957; IGY Report 21, p. 558.

[13] See, for example, ltrs, E. A. Roberts, Amateur Radio Station W2PEQ, to NRL, 6 Oct 1957, and Paul E. Smay, station W9TZN, to Dir/NRL, 6 Oct 1957; NYT, 6 Oct 1957.

[14] Hayes, Smithsonian Tracking Program: 2, pp. 226, 333, 340, 341; min 17th mtg TPESP, 3 Apr 1958, and 20th mtg, 21 July 1959; interview, Whipple.

[15] Hayes, *op. cit.*, p. 333.

[16] *Ibid.*, p. 343.

[17] *Ibid.*, p. 344; phone interview with E. Nelson Hayes, 30 Oct 1967; min 15th mtg TPESP, 7 Jan 1958.

[18] NASA Chronology, appendix A; see also ltr 4100–221, Whipple to J. Paul Walsh, 9 Oct 1957 [forwarding world map predictions for the Soviet satellite]; Johnson Subcommittee Hearings 1957, pt. 1, pp. 155f.

[19] Johnson Subcommittee Hearings, pt. 1, p. 2; ltr OP–515C Ser: 0278P51, Chief/Naval Operations to Chief/Naval Research, sub: Chronological History of Project Vanguard and Associated Preceding Projects, Request for; memo for the Sec/Navy . . . , 9 Oct 1957, sub: Briefing of Preparedness Investigating Subcommittee Staff on the Vanguard program . . . ; phone interview, Hagen, 30 Oct 1967; John B. Medaris and Arthur Gordon, *Countdown for Decision* (New York: G.P. Putnam's Sons, 1960), p. 135. For an amusing account of the preparation of the chronological history of Vanguard, see Stehling, *Project Vanguard*, pp. 95–97; NYT, 6 Jan 1958.

[20] Phone interview, Hagen; memo, Hagen to Codes 2000 and 1000, 15 Oct 1957, sub: TID Staff, Commendation of: Hagen, "The Viking and the Vanguard," in *History of Rocket Technology*, p. 137.

[21] House of Representatives, Committee on Appropriations, Subcommittee on Department of Defense Appropriations, *Hearings on Department of Defense Appropriations for 1960*, part 6, 86th Cong., 1st Sess., 14 Apr 1959, pp. 62f; memo, Chief/Naval Research to Chief/Naval Operations, 29 Oct 1957, sub: Project Vanguard Status Rpt No. 80.

[22] Quoted in Medaris and Gordon, *op. cit.*, p. 166.

[23] Ernst Stuhlinger, "Army Activities in Space —A History," *IRE Transactions in Military Electronics*, vol. MIL–4, Nos. 2–3 (Apr–July 1960), pp. 65f; min 14th mtg TPESP, 1 May 1957; ltr, Hagen to authors, 9 Jan 1968.

[24] Stuhlinger, *op. cit.*, Medaris and Gordon, *op. cit.*, pp. 119–20, 122, 134–36, 147, 151; NASA Chronology, p. 87; Eisenhower, *op. cit.* (n. 4), p. 210; see also conf rpt 4110–46, 18 Oct 1957, purpose: to Explore Capability of Jupiter Vehicles.

[25] Eisenhower, *op. cit.*, p. 244 n.; Medaris and Gordon, *op. cit.*, p. 155.

[26] Interviews with Siry, Rosen, and Berg, 21 Mar 1968; Medaris and Gordon, *op. cit.*, pp. 157–167; Eisenhower, *op. cit.*, p. 211; min 13th mtg TPESP, 22 Oct 1957.

[27] Interview, Eisenhower, 1966; Medaris and Gordon, *op. cit.*, p. 157. See also *JUNO* [a report], Pasadena, the Jet Propulsion Laboratory, no date; and Johnson Subcommittee Hearings 1957, pt. 1, pp. 207, 319, 374, 544f, 557, and 569.

[28] *Ibid.*, pp. 151, 162, 165f, and 318–320; see also memo, Sec/Navy to Sec/Defense, 31 Oct 1957, sub: Project Vanguard.

[29] Klawans and Burghardt, Vanguard Launch Vehicle, p. 58f; encl (1) ltr, Dir/NRL to Chief/Naval Research, 23 Dec 1957, sub: Preliminary Report on TV-3 (cited hereafter as Preliminary Report on TV-3), forwarding of; interviews, Bridger, 3 Aug and 13 Nov 1967; ltr, Vanguard Project engineer to BAR, Azusa, Cal, 25 July 1957, sub: Second Stage Vanguard, Status of; Burghardt Ms. Diary, 8–23–57; Weekly Progress Report 4104–47, Vanguard Planning Office to Vanguard Project Director, 3 Sept 1957; encl to memo for Mr. James O. Spriggs, Office Asst Sec/Def, 17 Sept 1957, sub: Draft of Proposed Report to the President, Submittal of; conf rpt 4120–498, 16 Sept 1957, purpose: Second-Stage Thrust Chamber Coat-

274

ings . . . ; memo, Chief/Naval Research to Chief/Naval Operations, 12 Sept 1957, sub: Project Vanguard Weekly Status Rpt No. 82; ltr 4120–558 Ser:01857, Dir/NRL to BAR, Azusa, 27 Nov 1957, sub: Second Stage Improvement Program, Decisions Concerning; encl (1) ltr, Dir/NRL to Chief/Naval Research, 27 Nov 1957, sub: Project Vanguard Weekly Status Rpt No. 92.

[30] See chapter 11, pp. 197–198.

[31] Klawans and Burghardt, Vanguard Launch Vehicle, p. 66; Preliminary Report on TV–3; NYT, 2 Dec 1957; Martin Monthly Progress Letter 18, p. 16; Vanguard Planning Office to Vanguard Project Dir, Weekly Progress Rpt of 4 Nov 1957; Martin Monthly Progress Letter 19, p. 9; encl (1) ltr, Dir/NRL to Chief/Naval Research, 19 Nov 1957, sub: Project Vanguard Weekly Status Rpt No. 91; interview, von Kármán, Sept 1961.

[32] NYT, 4 Dec 1957; Preliminary Report on TV–3.

[33] Ibid.; NYT, 6 Dec 1957.

[34] Preliminary Report on TV–3; NYT, 7 Dec 1957; Kurt Stehling, "Vanguard," in Arthur C. Clarke, ed., The Coming of the Space Age (see n. 1), pp. 15–20; Klawans and Burghardt, Vanguard Launch Vehicle, pp. 66f. Verses are from Robert Burns' "To a Mouse," and mileages at AFMTC are given in ltr, Commanding Office [PAFB] to Dir/NRL, 17 Jan 1958, sub: Mileage in Patrick-Cape Area, Forwarding of.

[35] Klawans and Burghardt, Vanguard Launch Vehicle, pp. 112f; phone interviews, 17 Nov 1967, with Richard Porter, Bernard Klawans, and N. J. Constantine; NYT, 7 Dec 1957; copy of memo in GE files, Klawans to Bridger, Burghardt and [Louis] Michelson, 15 Nov 1960, sub: Revision to Vanguard Flight Test History Summary; memo, Code 4101 to Codes 4100, 4110, 4120, 4160, 2 Jan 1958, sub: TV–3 Failure.

[36] NYT, 7 Dec 1967; interview, Markarian, 13 Oct 1967; Medaris and Gordon, op. cit. (see n. 19), p. 197; ltr 4100–340, Hagen to All Members of the Vanguard Staff, 18 Dec 1957, and enclosures. Bronk, speech at tenth anniversary Vanguard dinner, 16 Mar 1968.

Chapter 12

[1] For details concerning this transfer, see chapter 13, pp. 238–239.

[2] Lloyd V. Berkner, op. cit. (see ch. 7, n. 17), p. 478; Medaris and Gordon, op. cit. (see ch. 11, n. 19), pp. 167, 174, 190, 197f, and 200; Johnson Subcommittee Hearings, pp. 557–60 and 1704; memo, Chief/Naval Research to Chief/Naval Operations, 2 Jan 1957, sub: Project Vanguard Weekly Status Rpt No. 93; min 15th mtg TPESP, 7 Jan 1958; NYT, 1 Feb 1958; ltr, Thomas J. Killiam [Dept/Navy] to J. G. Reid, Jr. [National Academy of Sciences], 20 Feb 1958. For information on the Army's microlock system, see Corliss, The Evolution of STADAN, pp. 57 and 61; and W. K. Victor, H. L. Richter, and J. P. Eyraud, "Explorer Satellite Electronics," IRE Transactions on Military Electronics, vol. MIL–4, nos. 2–3 (April–July 1960), pp. 83–85.

[3] Vanguard Planning Officer to Vanguard Director, 17 Dec 1957, Weekly Progress Rpt; 2d endorsement on GLM ltr 0459 of 5 Mar 1958, Dir/NRL to Chief/Naval Research, 1 Apr 1958, sub: Contract Nonr 1817 (00), Project Vanguard Restoration of Launching Complex at AFMTC, Proposal for; memo, Dir/Navy Tests to Office, Information Services, Historical Branch, 13 Jan 1958, sub: Vanguard Project Case History Rpt, Dec 1957; Vanguard Planning Office to Vanguard Project Director, 17 Dec 1957, Weekly Progress Rpt; memo, Chief/Naval Research to Chief/Naval Operations, 23 Dec 1957, sub: Project Vanguard Weekly Status Rpt No. 92; Vanguard Planning Officer to Vanguard Project Director, 31 Dec 1957, Weekly Progress Rpt; memo, Chief/Naval Research to Chief/Naval Operations, 7 Jan 1958, sub: Project Vanguard Weekly Status Rpt No. 94; Medaris and Gordon, op. cit. (see ch. 11, n. 19), pp. 206f.

[4] Johnson Subcommittee Hearings, pp. 557f, 560, and 1715; Medaris and Gordon, op. cit. (see ch. 11, n. 19), pp. 197f, 200ff, and 218.

[5] Wernher von Braun, "The Explorers," in IXth International Astronautical Congress . . . 1958 Proceedings (Amsterdam, 1959), pp. 916–928; NYT, 1 Feb 1958; Medaris and Gordon, op. cit., pp. 204–226; min 16th mtg TPESP, 12 Feb 1958.

[6] See chapter 11, p. 207.

[7] Memo, Chief/Naval Research to Chief/Naval Operations, 15 Feb 1958, sub: Project Vanguard Weekly Status Rpt No. 98; Klawans and Burghardt, Vanguard Launch Vehicle, pp. 112f; Medaris and Gordon, op. cit., p. 229.

[8] Stehling, Project Vanguard, pp. 190 and

206; memo, Chief/Naval Research to Chief/Naval Operations, 25 Feb 1958, sub: Project Vanguard Weekly Status Rpt No. 98; Vanguard Planning Office to Vanguard Project Director, 24 Feb 1958, Weekly Progress Rpt; memo, Chief/Naval Research to Chief/Naval Operations, 28 Feb 1958, sub: Project Vanguard Weekly Status Rpt No. 100; Vanguard Planning Office to Vanguard Project Director, 4 Mar 1958, Weekly Progress Rpt; memos, Chief/Naval Research to Chief/Naval Operations, 7 Mar and 14 Apr 1958, sub: Project Vanguard Weekly Status Rpts Nos. 101 and 106.

[9] Min 17th mtg TPESP, 3 Apr 1958; Vanguard Planning Office to Vanguard Project Director, 17 Mar 1958, Weekly Progress Rpt; Stehling, *Project Vanguard*, pp. 211–215; NASA Chronology, p. 141; NYT, 18 Mar 1958; memo, Chief/Naval Research to Chief/Naval Operations, 21 Mar 1958, sub: Project Vanguard Weekly Status Rpt. No. 102; Klawans and Burghardt, Vanguard Launch Vehicle, pp. 114f; ltr, Dir/NRL to Dir/ARPA, 15 May 1958, sub: Scientific Satellite Program.

[10] Robert L. Rosholt, *An Administrative History of NASA, 1958–1963*, NASA SP–4101 (Washington, D.C.: NASA, 1966), pp. 6–40; Allison Griffith, *The National Aeronautics and Space Act: A Study of the Development of Public Policy* (Washington, D.C.: Public Affairs Press, 1962), pp. 9–16; Dwight D. Eisenhower, *op. cit.* (ch. 11, n. 4), pp. 257f; NYT, 8 Feb 1958; Medaris and Gordon, *op. cit.* (ch. 11, n. 19), p. 227.

[11] Stuhlinger, *op. cit.* (ch. 11, n. 23), pp. 68f; Medaris and Gordon, *op. cit.*, pp. 240f.

[12] Encl (1), ltr, Chief/Naval Research to Comptroller of the Navy, 26 May 1958, sub: Statement on Current Status of Project Vanguard dated 23 May 1958; ltr, GLM to Chief/Naval Research, 16 May 1958, sub: Contract Nonr–1817(00), Project Vanguard Modification of TV–4BU to Provide for Satellite Capability; ltr, GLM to Dir/NRL, 1 May 1958, sub: Contract Nonr–1817(00), Project Vanguard Employment of High Performance ABL Motor in TV–4BU; Vanguard Planning Office to Vanguard Project Director, 12 Apr 1958, Weekly Progress Rpt; memo, Chief/Naval Research to Chief/Naval Operations, 7 Apr 1958, sub: Project Vanguard Weekly Status Rpt, No. 103; Memo, Chief/Naval Research to Chief/Naval Operations, sub: Weekly Status Rpt. No. 104; ltr, J. Paul Walsh to Hon. [U.S.

Rep] David Dennison, 12 May 1958; ltr, T. Keith Glennan to John P. Hagen, 29 Sept 1958.

[13] Memo, Chief/Naval Research to Chief/Naval Operations, 1 May 1958, sub: Project Vanguard Weekly Status Rpt. No. 107; *ibid.*, 14 May 1958, Weekly Status Rpt. No. 108; ltr, Dir/NRL to GLM, 11 June 1958, sub: Contract Nonr 1817 (00), Project Vanguard, Correction of Deficiency; Klawans and Burghardt, Vanguard Launch Vehicle, pp. 114f. For dates and details on the first three mission-vehicle launches, see Appendix I.

[14] Klawans and Burghardt, Vanguard Launch Vehicle, pp. 116f; interviews, Homer E. Newell, 21 Dec 1967, and Klawans, 9 Jan 1968.

[15] Vanguard Planning Office to Vanguard Project Director, 25 Mar 1958, Weekly Progress Report; Klawans and Burghardt, Vanguard Launch Vehicle, pp. 20 and 116f.

[16] 8 December 1957.

[17] Public Opinion File, NHF; interview, Hastings and Harloff, 5 June 1968.

[18] The tenth anniversary dinner, a very special occasion, drew 250 veterans of the program, including some members of the media.

Chapter 13

[1] Min 14th mtg TPESP, 6 Nov 1957, pp. 7–8.

[2] See chapter 11, p. 195.

[3] See chapter 6, p. 107.

[4] Min 7th mtg WGII, 21 Oct 1957; min 13th mtg TPESP, 22 Oct 1957, pp. 2–6.

[5] See chapter 7, p. 124.

[6] Min 14th mtg TPESP, 6 Nov 1957.

[7] Associated Press releases, 17 Nov 1967.

[8] Min 14th mtg TPESP, 6 Nov 1957, pp. 2–6; and min 16th mtg TPESP, 12 Feb 1958, p. 9; interview, John Townsend, 12 Mar 1968.

[9] Min 15th mtg TPESP, 8 Jan 1958, pp. 2–4, 9, and min 16th mtg TPESP 12 Feb 1958, p. 4; IGY Report 21, pp. 560–61.

[10] See n. 4 and chapter 7, pp. 128–129; min 15th mtg TPESP, 8 Jan 1958, pp. 7, 9–10; attachment E to min 14th mtg USNC, 16 Jun 1958.

[11] Min 11th mtg USNC, 8 Jan 1958, pp. 1–2, and attachment A, ltr, John A. Simpson to Odishaw, 6 Jan 1958; min 26 mtg USNC–IGY Executive Committee, 8 Jan 1958, pp. 1–3.

[12] See chapter 6, p. 102; IGY Report 21, p. 561.

[13] Min 17th mtg TPESP, 3 Apr 1958, pp. 1–2, 7.

[14] Min 14th mtg USNC, 16 Jun 1958, p. 2; IGY Report 21, pp. 600–603.

[15] Min 14th mtg USNC, 16 June 1958, p. 1.

[16] IGY Report 21, pp. 593–597; "Symposium on Scientific Effects of Artificially Introduced Radiation at High Altitudes," *Proceedings of the National Academy*, vol. 45, no. 8 (15 Aug 1959).

[17] IGY Report 21, p. 566; min 14th mtg USNC, 16 June 1958, appendix B, Summary of Oral Report on the Earth Satellite Program by Richard W. Porter, pp. 1–2, and min 15th mtg USNC, 23 Oct 1958, attachment 6, pp. 6–11, 15–16, 19–21, 28.

[18] Memo, Hagen for James E. Webb, NASA Administrator, 2 Mar 1962, sub: Vanguard I, and enclosure, ltr, Hagen to Rear Admiral Coates, Chief Naval Research, 2 Mar 1961, NHF.

[19] *Astronautics and Aeronautics, 1965*, NASA SP-4006 (Washington, D.C.: NASA, 1966) p. 67.

[20] IGY Report 21, pp. 574–585, 592.

[21] See n. 14.

[22] Min 14th mtg USNC, 16 June 1958, pp. 1–2, 7, 10, and min 15 mtg USNC, 23 Oct 1958, pp. 1–3, 5–6, and attachment G; IGY Report 21, pp. 563–564; interview George Derbyshire IGY secretariat, 28 Nov 1967.

[23] See Robert L. Rosholt, *op. cit.* (ch. 12, n. 10) pp. 44ff.; min 20th mtg TPESP, 21 July 1959.

[24] See chapter 7.

[25] IGY Report 21, pp. 566, 604–605; Stuhlinger *op. cit.* (see ch. 11, n. 23), p. 69.

[26] The Advanced Research Projects Agency of the Defense Department and later NASA both sponsored a number of satellite flights during 1958 and 1959, eight of which were successful, but as none of these came under the aegis of the IGY National Committee and technical panels and all of them were subject to security classification that interfered with the release of information to other nations, they did not count as IGY satellites at all.

[27] IGY Report 21, pp. 607–608, 625.

[28] *Ibid.*, pp. 611–614.

[29] *Ibid.*, p. 629; Stuhlinger, *op. cit.* (see ch. 11, n. 23), p. 69; conf rpt, 3 Apr 1958, NRL master file, 4130–34; min 19th mtg TPESP, 17 Jul 1958, pp. 2–3; Satellite Experiments file folder 32.32, NAS Archives: IGY USNC.

[30] IGY Report 21, pp. 594, 616, 625.

[31] *Ibid.*, pp. 609, 621–623, 629.

[32] *Ibid.*, pp. 594–595.

[33] *Ibid.*, pp. 616–620.

Chapter 14

[1] Of these nineteen, three were Sputniks, one was a U.S.S.R. Lunik, planned as a lunar probe, and eight were American military satellites developed by the military outside the framework of the IGY. See chapter 13, n. 26.

[2] Klawans and Burghardt, Vanguard Launch Vehicle, pp. 7, 171–175.

[3] Ltr, Rosen to C. McL. Green, 15 Mar 1968, with encl, "A Brief History of Delta and Its Relation to Vanguard."

[4] Thomas E. Jenkins, "Budget Forecasting and Cost Control of Cost-Plus Development Contracts," 26 Oct 1959, copy in NHF.

[5] Interviews, Homer Newell, 19 Dec, and Capt. Berg, 28 Aug 1967; see Klawans and Burghardt, Vanguard Launch Vehicle, pp. 7, 171–175.

[6] Interview, Homer Newell, 26 Aug 1966.

[7] Min 19th mtg TPESP, 26 July 1958, p. 5.

SELECTED BIBLIOGRAPHY

Berkner, Lloyd V., editor. "Manual on Rockets and Satellites," volume 6 of *Annals of the International Geophysical Year*. New York: Pergamon Press, 1958. (Cited as: Manual on Rockets and Satellites.)

Clark, Arthur C., editor. *The Coming of the Space Age: Famous Accounts of Man's Probing of the Universe*. New York: Meredith Press, 1967.

Corliss, William R., "The Evolution of the Satellite Tracking and Data Acquisition Network (STADAN)." Goddard Historical Note No. 3. Greenbelt, Md.: Goddard Space Flight Center, 1967. (Cited as: The Evolution of STADAN.)

Eisenhower, Dwight D. *The White House Years: Waging Peace, 1956–1961*. New York: Doubleday & Company, Inc., 1965.

Emme, Eugene M. *Aeronautics and Astronautics: An American Chronology of Science and Technology in the Exploration of Space, 1915–1960*. Washington, D.C.: NASA, 1961. (Cited as: NASA Chronology.)

———, editor. *The History of Rocket Technology: Essays on Research, Development, and Utility*. Detroit: Wayne State University Press, 1964. (Cited as: *History of Rocket Technology*.)

Furnas, Clifford C. "Why Vanguard?" *Life*, volume 43, number 17 (2 October 1957), pages 22–25,

Goddard, Esther, and G. Edward Pendray, editors. *The Papers of Robert H. Goddard*. 3 volumes. New York: McGraw-Hill Book Co., 1969.

Griffith, Alison E. *The National Aeronautics and Space Act: A Study of the Development of Public Policy*. Washington: Public Affairs Press, 1966.

Hall, R. Cargill. "Vanguard and Orbiter." *The Airpower Historian*, volume 9, number 4 (October 1964).

Hayes, E. Nelson. "The Smithsonian's Satellite-Tracking Program: Its History and Organization [Part 1]," in *Annual Report . . . Smithsonian Institution . . . 1961*, pages 275–322, 4 plates, 7 figures. (Cited as: Smithsonian Tracking Program.)

———. "The Smithsonian's Satellite-Tracking Program: Its History and Organization, Part 2," in *Annual Report . . . Smithsonian Institution . . . 1963*, pages 331–357.

———. "Tracking Sputnik I," in *The Coming of the Space Age*, Arthur C. Clarke, editor. New York: Meredith Press, 1967.

Jones, Sir Harold Spencer. "The Inception and Development of The International Geophysical Year," in volume 1 (History of the IGY) of *Annals of the International Geophysical Year*. New York: Pergamon Press, 1959.

Klawans, B., and Joseph E. Burghardt. "The Vanguard Launching Vehicle: An Engineering Summary." Martin Engineering Report Number 11022, April 1960. (Cited as: Vanguard Launch Vehicle.)

Medaris, John G., and Arthur Gordon. *Countdown for Decision*. New York: G. P. Putnam's Sons, 1960.

Mengel, John T. "Tracking the Earth Satellite, and Data Transmission by Radio." *Proceedings of the IRE*, volume 44, number 6 (June 1956).

————, and Paul Herget. "Tracking Satellites by Radio." *Scientific American*, volume 198, number 1 (January 1958), pages 23–29.

National Academy of Sciences–National Research Council. "Report on the U.S. Program for the International Geophysical Year." IGY General Report Number 21, November 1965. (Cited as: IGY Report 21.)

Rosen, Milton. *The Viking Rocket Story*. New York: Harper & Bros., 1955.

Rosholt, Robert L. *An Administrative History of NASA, 1958–1963*. Washington: NASA, 1966.

Smitherman, W. E. "Army Participation in Project Vanguard." *IRE Transactions on Military Electronics*, volume MIL–4, numbers 2–3 (April–July 1960).

Stehling, Kurt. K. *Project Vanguard*. Garden City, N. Y.: Doubleday & Co., Inc., 1961.

Stuhlinger, Ernst. "Army Activities in Space—A History." *IRE Transactions on Military Electronics*, volume MIL–4, numbers 2–3 (April–July 1960).

Sullivan, Walter. *Assault on the Unknown: The International Geophysical Year*. New York: McGraw-Hill Book Co., 1961.

U.S. Congress, Senate. *Hearings Before the Preparedness Investigating Subcommittee of the Committee on Armed Services*, 85th Congress, 1st and 2d sessions, Parts 1 and 2. (Cited as: Johnson Subcommittee Hearings, 1957.)

Van Allen, James A., editor. *Scientific Uses of Earth Satellites*. Ann Arbor: University of Michigan Press, May 1956.

Victor, W. K., H. L. Richter, and J. P. Eyraud. "Explorer Satellite Electronics." *IRE Transactions on Military Electronics*, volume MIL–4, numbers 2–3 (April–July 1960).

Von Braun, Wernher. "The Explorers," in *IXth International Astronautical Congress, 1958 Proceedings*. Amsterdam: 1959.

Witkin, Richard, editor. *The Challenge of the Sputniks*. Garden City, N. Y.: Doubleday & Co., Inc., 1958.

APPENDIXES

1 Vanguard Flight Summary

2 Explorer Flight Summary

3 IGY Satellite Launches

1. VANGUARD FLIGHT SUMMARY

Vehicle (Launch Date)	Objectives	Results
TV-0 (8 Dec 1956) Viking No. 13, a liquid-propellant single-stage rocket redesignated TV-0 and fired as the first Vanguard test vehicle.	*Primary*: to evaluate the performance of the internal telemetry system, to evaluate the launch complex, and to become familiar with the operations, range safety, and tracking systems of the AFMTC rocket range. *Secondary*: to test the Vanguard Minitrack transmitter and to evaluate the coasting-flight attitude control system.	All objectives met except evaluation of coasting-flight attitude control. During powered flight, the performance of all components was either satisfactory or superior. Vehicle reached an altitude of 126.5 mi and range of 97.6 mi. Rocket-borne instrumentation and telemetry systems performed excellently; ground instrumentation coverage adequate.
TV-1 (1 May 1957) Two-stage test vénicle. 1st stage, the Viking no. 14 slightly modified for Vanguard objectives. 2d stage, a prototype solid-propellant Vanguard 3d stage. 2d-stage payload, an instrumented nosecone.	*Primary*: to flight-test the Vanguard 3d-stage prototype for spinup, separation, ignition, and propulsion and trajectory performance. *Secondary*: further evaluation of ground-handling procedures, techniques, and equipment and inflight instrumentation and equipment.	All test objectives met. Flight operation and performance of all powerplant systems very good. The vehicle was properly controlled throughout flight to an altitude of 121 mi and range of 451 mi.

TV-2
(23 Oct 1957)
Vanguard prototype consisting of a live 1st stage, a simulated but inert 2d stage, and an inert 3d stage.

Primary: to evaluate the Vanguard launch system and the flight performance of the 1st-stage propulsion system, the 2d-stage retrorocket system, and the 3d-stage spinup system and to obtain data on the 1st- and 2d-stage structural characteristics.

Secondary: to evaluate equipment, test, procedures, 1st-stage handling, and the SHF (C-band) beacon and radar equipment.

All test objectives met. Performance of all components throughout flight superior. This test confirmed that the 1st stage operated properly at altitude, conditions were favorable for successful separation of 1st and 2d stages, launch stand clearance for the condition of low surface winds was no problem, there was structural integrity throughout flight. Test also demonstrated dynamic compatibility between control system and structure.

TV-3
(6 Dec 1957)
First complete Vanguard test vehicle with three live stages.

Primary: to launch into orbit a minimal (6.4-in., 4-lb) satellite to determine atmospheric density and the shape of the earth. To evaluate satellite thermal design parameters and to check the life of solar cells in orbit.

Secondary: to test and evaluate all stages and systems of the vehicle. This was to have been the first flight firing of the 2d-stage propulsion system and of the complete Vanguard guidance and control system.

Less than one sec after liftoff, the 1st-stage engine lost thrust because of improper engine start. Vehicle settled back on launch stand and exploded.

TV-3BU
(5 Feb 1958)
Identical to TV-3.

Same as those of TV-3.

After 57 sec of normal flight, a control system malfunction caused loss of vehicle attitude control. Vehicle broke up only after an angle of attack of at least 45° had been exceeded.

Appendix 1—Continued
Vanguard Flight Summary

Vehicle (Launch Date)	Objectives	Results
TV-4 (17 Mar 1958) Identical to TV-3BU.	Same as those of TV-3BU.	Placed *Vanguard I*, totaling 57 lb (a 3.25-lb payload and the 53-lb 3d-stage motor case), in orbit originally expected to last up to 2,000 years (later estimate, 240 years). Initial orbit had apogee of 2,465 mi, perigee of 406 mi, and period of 134 min. Guidance system produced an overall error of less than 1° in satellite injection angle. First use of solar cells to supply power for satellite instruments.
TV-5 (28 Apr 1958) Final test vehicle, differing from a production satellite launching vehicle (SLV) only in the greater degree of instrumentation.	*Primary*: to launch into orbit a fully instrumented 20-in., 21.5-lb "x-ray and environmental" satellite. This satellite was to study maximum variations in the intensity of solar x-ray radiation in the 1 to 8 Å wavelength bands and to make certain space environment measurements. *Secondary*: to verify the performance of the complete vehicle.	Flight normal through 2d-stage burnout, but 2d-stage shutdown sequence was not completed electrically, which prevented arming of the coasting-flight control system and separation and firing of 3d stage. 2d-stage performance below nominal, but combined 1st- and 2d-stage performance somewhat better than nominal.
SLV-1 (27 May 1958) First production satellite launching vehicle.	To launch into orbit a fully instrumented, 20-in., 21.5-lb Lyman-alpha satellite. This satellite was to study solar Lyman-alpha radiation and to make certain space environment measurements; it was identical to the	Successful operation and performance achieved throughout flight, except at 2d-stage burnout. At that time, a disturbance caused loss of attitude reference to the pitch gyro so that the remainder of the flight was

Vehicle (date)	Objective	Results
	x-ray satellite of TV–5 except that it covered the 1100 to 1300 A wavelength bands.	controlled to a false reference. Third stage launched at an angle of approximately 63° to the horizontal, thus precluding a satisfactory orbit.
SLV–2 (26 June 1958)	Same as the primary objectives of TV–5.	Second-stage propulsion system shut down after 8 sec of burning, so that the velocity was low and the 3d stage was never armed for firing. As a normal result of the premature shutdown, 2d-stage propellant tank pressures exceeded design values, proving the structural integrity of the tankage.
SLV–3 (26 Sep 1958)	To launch into orbit a 20-in., 23.3-lb "cloud cover" satellite. This satellite was to measure the global distribution and movement of cloud cover and to contribute to the basic knowledge of the earth's energy budget.	Flight normal (or better) in all respects, except that 2d-stage performance was well below minimum predicted. Burned-out 3d stage and satellite reached an altitude of about 265 mi, but the velocity was about 250 fps short of the 25,000 fps required to orbit. The satellite was presumably destroyed during atmospheric reentry some 9,200 mi downrange.
SLV–4 (17 Feb 1959)	To launch into orbit a 20-in., 23.7-lb "cloud cover" satellite practically identical to that of SLV–3.	Placed *Vanguard II*, totaling 71.5 lb (23.7-lb payload and 47.08-lb 3d-stage motor case), in an orbit expected to last at least 200 years. Initial orbit had apogee of 2,063 mi, perigee of 346 mi, and period of 125.9 min. Guidance system produced a negligible overall error in injection angle of $0.02° \pm 0.2°$.

Vehicle (Launch Date)	Objectives	Results
SLV-5 (13 Apr 1959)	To launch into orbit a fully instrumented 13-in. diameter magnetometer satellite and an expandable (30-in.) aluminum sphere. The satellite was to determine if the predicted Stormer-Chapman ring current existed and to improve knowledge of the earth's magnetic field. The expandable sphere was to supply information on upper air density.	Pitch attitude control of 2d-stage lost during 1st-stage separation. Resulting tumbling motion in the pitch plane aborted the flight.
SLV-6 (22 June 1959)	To launch a 20-in. diameter, 23.8-lb "radiation balance" satellite into an orbit with a relatively high inclination (about 48°) to the equator. This satellite was to measure the direct radiation of the sun, the radiation reflected from the earth, and the long-wave radiation emitted by the earth and its atmosphere.	There was a rapid decay of tank pressures immediately after 2d-stage ignition. Abnormally low flow rates and chamber pressures resulted, accompanied by combustion instability. About 40 sec later, the helium sphere exploded from unrelieved buildup of pressure by the heat generator. The trajectory was accurately modified from a launch azimuth of 100° to a flight azimuth of about 48° by use of inflight roll programming just after launch.

TV–4BU
(18 Sep 1959)

This vehicle incorporated the Allegany Ballistics Laboratory X248 A2 solid-propellant motor as the 3d stage in place of the Grand Central motor used in previous Vanguard vehicles.

To launch into orbit a fully instrumented 52-lb "magnetometer, x-ray, and environmental" satellite. This payload combined the scientific objectives of the TV–5 and SLV–5 satellites.

Placed *Vanguard III*, totaling 94.6 lb (52.25-lb payload and 42.3-lb 3d-stage motor case), in an orbit expected to last at least 50 years. Initial orbit had apogee of 2,326 mi, perigee of 317 mi, and period of 130 min. Guidance system produced a negligible overall error in injection angle of 0.05° ± 0.2°.

2. EXPLORER FLIGHT SUMMARY

Vehicle (Launch Date)	Objectives	Results
Jupiter C (31 Jan 1958)	To launch into orbit an 18.13-lb "cosmic ray" satellite.	Placed *Explorer I* (totaling 30.8 lb, including the 18.13-lb satellite) in an orbit expected to last 3-5 years. Initial orbit had apogee of 1,573 mi, perigee of 224 mi, and period of 114 min. Scientific instrumentation confirmed predicted cosmic radiation levels up to 600-mi altitude, measured frequency of impacts and size of micrometeoroids, provided information on earth's bulge and gravity, obtained data on atmospheric density at extreme altitude, and confirmed success of method to control temperature of satellite's interior. Count rates above 600-mi altitude were very irregular, later found to be the result of very intense trapped radiation about the earth.
Jupiter C (5 Mar 1958)	To launch into orbit an 18.5-lb "cosmic ray" satellite similar to *Explorer I* and known as Explorer II.	Failed to orbit. Last stage did not ignite. Flight time 885 sec.
Jupiter C (26 Mar 1958)	To launch into orbit a payload almost identical to Explorer II.	Placed *Explorer III* in an orbit that lasted until 28 June 1958. Initial orbit had apogee of 1,740 mi, perigee of 119 mi, and period of 115.8 min. Scientific instrumentation added to data acquired by *Explorer I*.

Vehicle (date)	Purpose	Result
Jupiter C (26 July 1958)	To launch into orbit a 25.8-lb payload carrying two Geiger-Mueller counters, two scintillation counters, and internal temperature measurements transmitted by subcarrier center frequency shift.	Placed *Explorer IV* in orbit. Initial orbit had apogee of 1,373 mi, perigee of 163 mi, and period of 110 min. Measured cosmic rays and trapped radiation over a wide range of levels and energies. Also explored a far greater volume of space as regards latitude and altitude than *Explorers I* and *III*. Collected data on trapped electrons resulting from Argus high-altitude nuclear explosions.
Jupiter C (24 Aug 1958)	To launch into orbit a duplicate of *Explorer IV* to be known as Explorer V.	Failed to orbit. At 1st-stage separation, the booster rammed the instrument compartment. Flight time, 659 sec.
Jupiter C with "apogee kick" 5th stage added. (22 Oct 1958)	To launch into orbit a high-visibility balloon (9.26-lb, 12-ft diameter when inflated), to provide high-altitude atmospheric-density data and to serve as a radar target.	Failed to orbit. Rotational spin vibrations of the cluster caused the payload to drop off at 112 sec.
Juno II (16 July 1959)	To launch into orbit a multiple-experiment satellite and double-truncated cone (91.5 lb).	Failed to orbit. At liftoff the vehicle deviated sharply to the left and was destroyed at 5½ sec after liftoff. Failure of the guidance inverter caused open loop drift of the control system.
Juno II (14 Aug 1959)	Same as attempt of 22 Oct 1958.	Failed to orbit.
Juno II (13 Oct 1959)	Same as attempt of 16 July 1959.	Placed *Explorer VII* in orbit. Initial orbit had apogee of 681 mi, perigee of 345 mi, and period of 101.3 min.

3. IGY SATELLITE LAUNCHES

[United States IGY launches in boldface. Satellites and probes other than United States IGY launches have been inserted to complete the 1957–1959 chronological listing.]

Name (Launch date)	Experiments	Down	Inclination (Degrees)	Initial Perigee [1] (Miles)	Initial Apogee [1] (Miles)
1957					
Sputnik I [2] 1957 Alpha (4 Oct 1957)					
Sputnik II [2] 1957 Beta (3 Nov 1957)					
1958					
Explorer I 1958 Alpha (31 Jan 1958)	Cosmic rays Meteoroid erosion Temperatures	In orbit	33.3	224	1,573
Explorer II (5 Mar 1958)	Meteoric dust Meteoroids Temperatures Cosmic rays	Failed to orbit			
Vanguard I 1958 Beta (17 Mar 1958)	Temperature	In orbit	34	406	2,465
Explorer III 1958 Gamma (26 Mar 1958)	Cosmic rays Meteoroid erosion Temperatures	28 Jun 1958	33.4	119	1,740
Vanguard Test Vehicle 5 (28 Apr 1958)	Solar x-rays Environmental	Failed to orbit			
Sputnik III [2] 1958 Delta (15 May 1958)					
Vanguard SLV–1 (27 May 1958)	Lyman-alpha Environmental	Failed to orbit			

Name (Launch date)	Experiments	Down	Inclination (Degrees)	Initial Perigee [1] (Miles)	Initial Apogee [1] (Miles)
Vanguard SLV–2 (26 Jun 1958)	Solar x-rays Environmental	Failed to orbit			
Explorer IV 1958 Epsilon (26 Jul 1958)	Trapped radiation	23 Oct 1958	50.3	163	1,373
Explorer V (24 Aug 1958)	Trapped radiation	Failed to orbit			
Vanguard SLV–3 (26 Sept 1958)	Cloud cover	Failed to orbit			
Pioneer I 1958 Eta 1 (11 Oct 1958)					
Explorer VI [3] (22 Oct 1958)	12-ft inflatable sphere	Failed to orbit			
Pioneer II (at 963 mi, 3d stage failed to ignite) (8 Nov 1958)					
Pioneer III 1958 Theta (6 Dec 1958)					
Project Score 1958 Zeta (18 Dec 1958)					
1959					
Luna I [2] 1959 Mu 1 (2 Jan 1959)					
Vanguard II 1959 Alpha (17 Feb 1959)	Cloud cover Internal temperature	In orbit	32.8	346	2,063
Discoverer I 1959 Beta (28 Feb 1959)					

IGY Satellite Launches

Name (Launch date)	Experiments	Down	Inclination (Degrees)	Initial Perigee [1] (Miles)	Initial Apogee [1] (Miles)
Pioneer IV 1959 Nu 1 (3 Mar 1959)					
Discoverer II 1959 Gamma (13 Apr 1959)					
Vanguard SLV–5 (13 Apr 1959)	30-in. inflatable sphere Magnetometer	Failed to orbit			
Vanguard SLV–6 (22 Jun 1959)	Earth energy balance	Failed to orbit			
Explorer (16 Jul 1959)	Duplicate of *Explorer VII*	Failed to orbit			
Explorer VI [3] 1959 Delta (7 Aug 1959)					
Discoverer V 1959 Epsilon (13 Aug 1959)					
Explorer (14 Aug 1959)	Duplicate of Explorer VI	Failed to orbit			
Discoverer VI 1959 Zeta (19 Aug 1959)					
Luna II [2] 1959 Xi 1 (12 Sept 1959)					
Vanguard III 1959 Eta (18 Sept 1959)	Magnetometer Solar x-ray Lyman-alpha Environmental	In orbit	33.3	317	2,326

Name (Launch date)	Experiments	Down	Inclination (Degrees)	Initial Perigee [1] (Miles)	Initial Apogee [1] (Miles)
Luna III [2] 1959 Theta (4 Oct 1959)					
Explorer VII 1959 Iota (13 Oct 1959)	Micrometeoroid Cosmic rays Heavy nuclei Earth energy balance Solar x-ray Lyman-alpha Radio signals Ground studies	In orbit	50.3	345	681
Discoverer VII 1959 Kappa (7 Nov 1959)					
Discoverer VIII 1959 Lambda (20 Nov 1959)					

[1] Orbit figures are approximate.   [2] U.S.S.R.

[3] Two payloads were named Explorer VI: an IGY payload that failed to orbit 22 Oct 1958 and a post-IGY experiment launched successfully 7 Aug 1959.

INDEX

C

C-band, 254, 283
Cadiz, Spain, 150
Cadmium, 138
Cadmium-sulphide photosensitive cell, 247
Cal Tech. See California Institute of Technology.
Calhoun, Cdr. Howard W. (USN), 166
California, 7, 8, 122, 194, 212, 217, 219
California Institute of Technology (Cal Tech), 7, 26, 35, 199, 213
California, University of, Los Angeles (UCLA), 14
Cambridge, Massachusetts, 99, 154, 190, 194
Camera, 27, 149–154, 240
Campbell, Mrs. Lillian M., 74 ill.
Cape Canaveral, Florida,
   facilities, 74, 82, 91, 131, 136, 165–168, 178, 213, 214
   location, 34, 74, 76, 87, 133, 134, 141, 155, 161, 162–163, 172, 177, 180, 185, 197, 204, 206–207, 213, 218, 228, 230, 234, 237
   tests, 50, 93, 131
   See also AFMTC.
Carbon, 114
Carnegie Institution, 19, 31
Carothers, Neil, 37
Case Institute of Technology, 222
Cell, yeast, 124
Central Control, 162, 163, 172
Central Intelligence Agency (CIA), 29, 31
Cerenkov detector, 114
Chamber, thrust, 86, 204
Chapman, Sydney, 18, 37
Charlotte, North Carolina, 140
Chesapeake Bay, 77
Chicago, Illinois, 207
Chicago, University of, 241
Chief of Naval Operations, 178
Chile, 158
Chromium, 127
Chrysler Corp., 53
CIA. See Central Intelligence Agency.
CIG (Comité Internationale Geophysique), 245
Cincinnati Observatory, 146, 159, 161
Cincinnati, University of, 73
Cinetheodolite, 195
Circulation, 249
Civil Air Patrol, 230
Clark University, 2
Clean room, 168
Clemence, Gerald M., 73, 98, 159
Clement, George H., 35, 36, 49
Cleveland, Ohio, 222
Cloud cover, 126, 239, 246, 249, 285, 291
Cocoa Beach, Florida, 133, 135, 166
College, Alaska, 194
Collier's, 16

Comité Internationale Geophysique. See CIG.
Comité Spéciale de l'Année Géophysique Internationale (CSAGI). See CSAGI.
Committee for Evaluating the Feasibility of Space Rocketry, 6
Committee on Space Flight, 14–15
Committee on Space Research (COSPAR), 246–247
Computer, 159, 162
   704, 108, 160, 162
   709, 161, 162
Congress of the International Astronautics Federation, 17
Congress, United States, 21, 30, 130, 131, 133, 222, 240, 245
   appropriations, 21, 36, 102
   Senate Committee, 131, 196, 203, 214
Connecticut, 125
Consolidated Vultee Aircraft Corp. See Convair.
Consultant, 159
Contract, 66
   Vanguard, 58–61, 64–68
Control system, 217, 283, 289
Convair (Consolidated Vultee Aircraft Corp.), 145
Copenhagen, Denmark, 39
Cornell, Douglas, 37, 38 ill.
Cornell University, 36
Corrective Action Team, 226
"Cosmic Debris of Interplanetary Space," 242
Cosmic dust, 247
Cosmic ray, 6, 11, 16, 21, 39, 46, 109, 114, 117, 118, 119, 123, 124, 196, 238, 240, 241, 242, 243, 248, 288, 289, 290, 293
Cosmos Club, 106
COSPAR. See Committee on Space Research.
Cost, 49, 110, 137, 255
   Atlas-B satellite, 41
   computer, 160
   optical tracking, 108–109, 149, 240
   Orbiter, 43
   radio tracking, 137, 241
   Vanguard, 62–63, 104–105, 130–131
   Viking-based, 53
Countdown, 170, 214, 215, 218
Counter, frequency, 192
Crane, Jim, 192
CSAGI (Comité Speciale de l'Année Geophysique Internationale), 22, 23, 29, 37, 39, 104, 109, 110, 131, 185–186, 190, 193–194, 238, 245
Cuba, 77, 158
Cumberland, Md., 90
Cunningham, C. B., 192
Curacao, Netherlands West Indies, 152

D

Dahlgren, Va., 69
Daniel and Florence Guggenheim Foundation, 2

297

N

NACA. See National Advisory Committee for Aeronautics.

Naini Tal, India, 150

NAS. See National Academy of Sciences.

NASA. See National Aeronautics and Space Administration.

NASA Historical Archives, 209

National Academy of Sciences (NAS), 12, 14, 55–56, 185, 212, 222, 242, 245
    funds, 97–99, 102–106, 108, 129, 130
    IGY, 21–22, 25, 28–30, 62, 67, 80, 98–99, 110, 129–130, 131, 193, 198, 201, 213, 247, 256
    relationships, 18, 20, 24, 35, 38, 106–108, 253

National Advisory Committee for Aeronautics (NACA), 122, 222

National Aeronautics and Space Act, 245

National Aeronautics and Space Administration (NASA), 90, 107, 222–223, 235, 237, 245, 254–255

National Aeronautics and Space Agency, 222

National Bureau of Standards, 21, 22, 240

National Institutes of Health, 124, 240

National Research Council, 13, 21, 35

National Science Foundation (NSF), 98, 219
    funds, 21, 36, 63, 73, 97–100, 102–103, 105, 107, 130–131, 203, 241
    organization, 12
    role, 12, 15, 18, 21, 25, 28–29, 34–35, 38, 107–108, 130, 153

National Security Council, 30, 31, 34, 36, 62–63, 105, 134, 252

National Telecommunications Research Center, 158

Navaho (missile), 9

Naval Electronic Laboratory. See Navy, U.S.

Naval Industrial Fund, 64

Naval Observatory, 73, 159

Naval Ordnance Research Center, 69

Navy, U.S., 1, 4, 7–9, 18, 25, 35, 51, 55, 66, 75, 83, 108, 130–131, 140–141, 239, 246, 253
    Bureau of Aeronautics, 6–8, 12, 17, 19, 92
    Bureau of Ordnance, 6
    Bureau of Yards and Docks, 155
    Naval Electronic Laboratory, 155
    Naval Observatory, 73, 159
    Naval Research Laboratory. See NRL.
    Naval School of Aviation Medicine, 15
    Office of Naval Research (ONR). See Office of Naval Research.

"Navy-Martin Vanguard Operations Group, The," 167

Nazi, 5, 48

Neff, Robert, 166

Neptune (sounding rocket), 10

Nesmeyanov, A.N., 23

New Haven, Conn., 194

New Mexico, 2, 5, 195

New Mexico College of Agriculture and Mechanical Arts, 84
    Physical Science Laboratory, 84

New York, 160, 196

New York *Herald Tribune,* 39

New York, N.Y., 14, 15, 37, 187, 210

New York Stock Exchange, 210

New York Times, 186, 206, 210

Newcomb, Simon, 4

Newell, Homer E., Jr., 14, 15–16, 22, 25, 27, 62, 73, 74 ill., 97, 104, 107, 108, 115, 117, 118, 125, 127, 128 ill., 131, 185, 192, 237, 239, 240, 250 ill., 256

Newsweek, 187

Newton, Philippe W., 20 ill.

Newton, Sir Isaac, 1

Nicolet, Marcel, 37

Nike (missile), 70

Nitric acid, red fuming, 80, 89

Nitric acid, white fuming, 80, 88, 89

Nitric oxide, 117

Nitrogen, 114, 180

Nixon, Vice President Richard M., 212

North America, 154, 155

North American Aviation, Inc., 7, 9, 12, 52
    Rocketdyne Division, 17

North American Weather Consultants, 15

Norwalk, Conn., 152

Nosecone, 87, 89, 209

Nozzle, 80, 86

NRL (Naval Research Laboratory), 36, 44, 53, 56, 84, 104, 136–137, 174, 192, 194, 195, 201, 206, 207, 219, 235, 246, 253–254
    Applications Research Div., 73
    Astronomy and Astrophysics Div., 62
    Atmosphere and Physics Div., 11, 26,
    contract, 57–78, 79–85, 88–94, 97, 118, 133, 134–136, 159–160, 255
    Electronics Div., 126
    Electron Optics Branch, 10, 117
    funds, 11–12, 104–105, 130–131
    Micron Waves Branch, 6
    operations, 165–170, 174–181, 247
    Optics Div., 6, 117, 126
    planning, 7, 18, 43, 45, 52, 53–54, 80, 82–83, 85–86, 104, 108, 118, 124, 148, 160–161, 255
    research, 6, 9, 115–117, 125–130, 163, 238–239, 244, 252, 255
    Rocket Development Branch, 26, 30, 61, 62
    Rocket Sonde Research Section, 6, 9–10, 26, 65
    role, 45, 55, 57, 84, 101, 106–107, 213, 223
    scheduling, 140–143, 160–161
    Security Review Branch, 230

NSF. See National Science Foundation.

Nuclei, heavy, 239, 248, 293

Nunn, Joseph, 152

O

Oberth, Hermann, 2, 4
Observatory Philharmonic Orchestra, 194
Odishaw, Hugh, 21, 22, 25, 28, 97, 103, 110, 118, 131, 250 ill.
Office of Naval Research (ONR), 6, 11, 18, 26, 43, 57, 59–60, 61, 63, 75, 88, 94, 108, 160
O'Hea, Maj. John T. (USA), 73
O'Keefe, Maj. John, 15
Olifansfontein, S. Africa, 150
ONR. See Office of Naval Research.
Operation Paper Clip, 5
Orbiter, Project, 18, 22, 25, 26, 27, 30, 33, 36, 41, 43, 46, 48–53, 55, 74, 148, 198–199, 252
Organ Pass, N. Mex., 150, 153
Oscillator, 123, 146, 158, 174, 192–193, 245
O'Sullivan, William J., 122, 123
Ottenstroer, H.W., "Ott," 230
Oxydizer, 17, 80, 89, 168
Oxygen, 114
Oxygen, liquid (lox), 17, 80, 86, 172, 180

P

Pacific Ocean, 109
Pad, launch, 139–140
PAFB. See Patrick Air Force Base.
Palomar Observatory, 15
Pampa de Ancon, Peru, 158
Pan American Airways (Pan Am), 134, 174, 208
Panama, 77, 158
Panel on Upper Atmosphere Rocket Research, 14
"Panic button," 162
Paramo de Cotopaxi, Ecuador, 158
Particle, charged, 251
Pasadena, Calif., 18, 36, 152, 194, 213
Patrick AFB, Fla., 51, 60, 61, 133–136, 140, 155, 161
Peavy, Ross, 104
Peenemuende, 5, 7, 16
Peldehune Military Reservation, Chile, 158
Per diem, 167–168
Perkin-Elmer Corp., 152, 153
Peru, 158
Peterson, Cdr. W. J., 74 ill.
Petri, George W., 20 ill.
Phosphorous 32, 115
Photocell, 120, 123
Photograph, 46, 208, 239
Photon counter, 10
Physics and Medicine of the Upper Atmosphere (symposium), 14
Pickering, William H., 14, 26, 97, 101, 109, 123, 124, 193, 216 ill., 238–239
Pierce, John R., 15
PIO (Public Information Officer), 228–235
Piore, Emmanuel R., 18
Pirani (gauge), 116

"Planetary and interplanetary investigations," 245
Plan, long-term, 245
Point Mugu, Calif., 33, 83
Poland, 188
Pole, geomagnetic, 16
Poloskov, Sergei M., 185–186
Pomerantz, Martin A., 124, 239, 248
Porter, Richard W., 15, 35, 97, 102, 103, 106, 107, 109, 110, 118, 124, 131–132, 152–153, 186, 193, 231 ill.
Poughkeepsie, N.Y., 160
PPM/AM (pulse-position-modulation/amplitude-modulated), 83–84
President's Scientific Advisory Committee (PSAC), 31
Press conference, 37–38
Press statements, 39, 165, 186–187, 198, 206, 232, 234, 239
Pressure, 72, 93, 109, 116, 196
Pressurization, 88, 93, 177–178
Princeton University, 98, 113
"Proposal for Cosmic Ray Observations in Earth Satellites, A," 113–115
Propulsion, 27, 70, 79–80, 87–88, 253–254
    hydrogen, 9
    hypergolic, 70
    solid, 72, 254
Proton, 122, 243, 248
PSAC. See President's Scientific Advisory Committee.
Public information officer. See PIO.
Pulse-position-modulation/amplitude-modulated (PPM/AM), 83–84
Pulse-width-modulation/frequency-modulation, PWM/FM, 82–84
Purdy, William G., 59, 64–65, 66
Putt, Gen. Donald, 5
PWM/FM (pulse-width-modulation/frequency-modulation), 82–84

Q

Quarles, Donald A., 30, 31–33, 34–35, 36–37, 48, 52, 63, 64, 75, 84
Quito, Ecuador, 158

R

Radar, 45, 46, 83, 131, 141, 161, 163, 283, 289
    directional, 43
Radiation, 114, 122–123, 126, 240, 242–244, 248–249, 286, 290
    corpuscular, 243
    solar, 10, 109, 114, 116, 123, 196, 239, 244, 246, 248, 249, 284, 286, 290, 291, 292, 293
    trapped, 247, 248, 288, 289, 291
Radiation balance, 122–123, 239, 249, 286
"Radiation Balance of the Earth," 122–123
Radiation, Inc., 108, 163
Radiation pressure, 244

radar, 43, 159
radio, 18, 26, 43, 66, 99, 106, 109, 145, 154–159, 185–186, 193, 208, 241
  requirements, 49, 91, 109, 146
  stations, 78, 109, 110, 185–186, 241
  See also Minitrack.
Trajectory, 69, 255
Transducer, 120
Transistor, 116, 120, 129
Transmitter, 120, 126, 129, 146, 155, 238, 240
  airborne, 146, 173–174, 185–186, 196, 198, 204, 209, 243, 247, 282
  ground, 193
Transponder, 146
Treadwell Construction Co., 138
Treasury, Department of the, 31
Trimble, George S., Jr., 207
Troposphere, 247
Truax, Cdr. Robert C., 20 ill.
Truman, President Harry S., 16, 189
TRX–217 (propellant), 72
Tsiolkovskiy, Konstantin, 2, 4, 23
Tsiolkovsky Gold Medal, 32
Tucker, Capt. Samuel (USN), 53, 58, 60, 61, 62, 67, 75
Tumble rate, 226
Tungsten carbide, 204
Turbine, 86, 181
Turbopump, 88, 210, 254
Tuve, Merle, 31
TV–0, 93, 144, 172
  launch, 172–174, 282
TV–1, 93, 142–143, 174–176
  launch, 176, 282
TV–2, 131, 176–183, 185, 197–198, 204–205, 209
  launch, 176–183, 283
TV–2BU, 182, 228
TV–3, 93, 143, 170, 197–198, 204, 206–212, 214, 217, 218, 232, 234, 256
  launch, 208–209, 283
  static test, 170
TV–3BU, 214, 215, 217, 218, 284
  launch, 217, 283
  static test, 170
TV–4, 93, 142–143, 218, 223, 226
  launch, 218–221, 284
TV–4BU, 223, 228, 234, 235
  launch, 228, 287
  See also Vanguard III.
TV–5, 93, 142, 204, 223, 287
  launch, 223, 284, 290

U

UDMH (unsymmetrical-dimethylhydrazine), 50–51, 54, 88, 254
Ultraviolet, 6, 10, 39, 46, 117, 196, 248, 249, 252

Umbilical, 172
  cooling-air, 172
  electrical, 172
Union of Soviet Socialist Republics (U.S.S.R.), 9, 22, 23, 32, 39, 52, 129, 182, 185–190, 196, 203, 210, 238, 243
  Academy of Sciences, 23, 32
  National Committee, 109
  Rocket and Earth-Satellite Program for the IGY, 129
  satellite, IGY, 109–110, 185
United Nations, 210
United States (U.S.), 5, 17, 20, 28, 32, 37, 52, 109, 132, 150, 154, 167, 185–187, 189, 190, 196, 201, 208, 210, 219, 232, 237, 241, 243, 252–253, 256
United States National Committee for the IGY (USNC), 21, 23, 25, 27–31, 33, 34, 36, 55–56, 97–107, 109, 110, 113, 124, 130, 152–153, 237–238, 240–241, 245
  Technical Panel on the Earth Satellite Program (TPESP), 89, 97–98, 101–108, 110–111, 115, 117, 119–120, 123–124, 130, 131, 152–153, 154, 237–241, 245–246, 247, 256
United States Weather Bureau, 122
University, 97, 241
Unsymmetrical-dimethylhydrazine. See UDMH.
Upper Atmosphere Rocket Research Panel, 6, 17, 26, 103
URSI. See International Union of Scientific Radio Unions.
USAF. See Air Force, U.S.
USN. See Navy, U.S.
USNC. See United States National Committee for the IGY.
U.S.S.R. See Union of Soviet Socialist Republics.
"U.S.S.R. Rocket and Earth-Satellite Program for the IGY," 129

V

V–2 (rocket), 4–6, 14, 17, 44
Valve, 182, 207
Van Allen, James A., 17, 18, 25, 89–90, 98, 103, 104, 107, 110, 113–115, 118, 119, 122, 123, 124, 128–129, 131, 216 ill., 239, 242, 247–248
Van Allen radiation belts, 243, 246–247, 248–249, 251
Vance, Cyrus, 214
Vanguard (rocket), 12, 114, 118, 143, 159, 172, 182, 224
  cost, 130–131
  flight schedule, 76, 88, 93, 94, 174, 213
  first stage, 79, 83, 86–87, 114, 142, 169, 170, 177–178, 182, 197, 209, 217, 226, 253–254, 255, 282, 283, 284